KT-172-568

THE STORY OF SCIENCE
POWER, PROOF AND PASSION

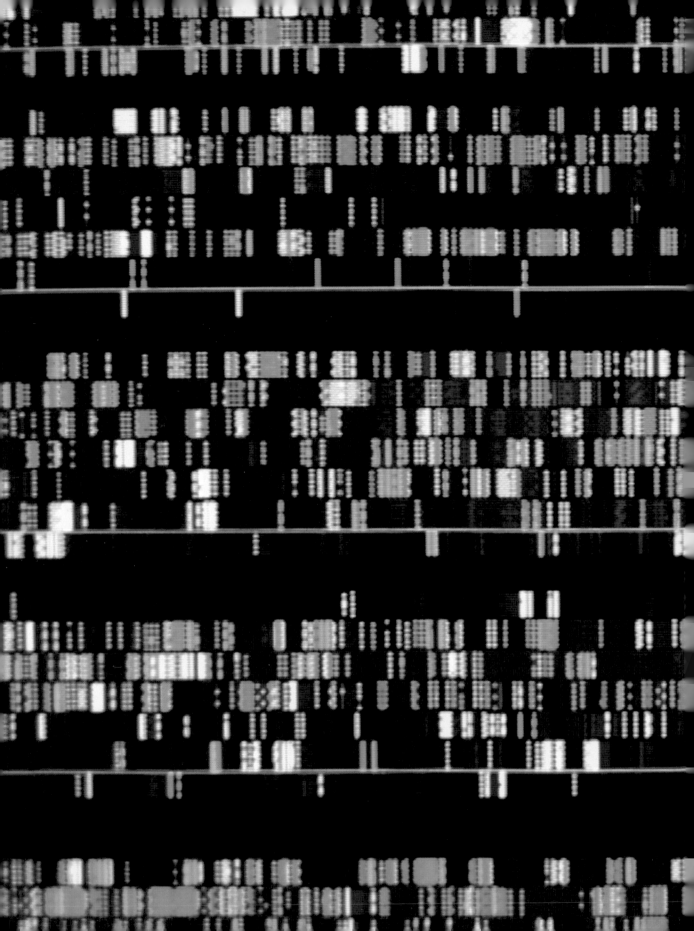

THE STORY OF SCIENCE
POWER, PROOF AND PASSION

MICHAEL MOSLEY AND JOHN LYNCH

Left:
Human DNA
represented as
a series of coloured
bands. The Human
Genome Project,
an international
effort to map the
genetic code of our
species, is one of
the most ambitious
scientific projects
ever undertaken.

MITCHELL BEAZLEY

The Story of Science: Power, Proof and Passion
Michael Mosley and John Lynch

First published in Great Britain in 2010 by Mitchell Beazley,
an imprint of Octopus Publishing Group Limited,
Endeavour House, 189 Shaftesbury Avenue
London, WC2H 8JG
www.octopusbooks.co.uk
An Hachette UK Company
www.hachette.co.uk

Copyright © Octopus Publishing Group 2010
Text copyright © Michael Mosley and John Lynch 2010

By arrangement with the BBC
The BBC logo is a trade mark of the British Broadcasting Corporation and is used under licence.
BBC logo © BBC 1996

All rights reserved. No part of this work may be reproduced or utilized in any form or by any
means, electronic or mechanical, including photocopying, recording or by any information
storage and retrieval system, without the prior written permission of the publisher.

While all reasonable care has been taken during the preparation of this editon, neither the publisher,
editors, nor the authors can accept responsibility for any consequences arising from the use thereof
or from the information contained therein.

ISBN: 978 1 84533 547 2

A CIP catalogue record for this book is available from the British Library.

Commissioning Editor: Peter Taylor
Art Director: Pene Parker
Deputy Art Director: Yasia Williams-Leedham
Designer: Mark Kan
Project Editor: Georgina Atsiaris
Copy Editor: Hayley Birch
Proofreader: Jo Murray
Indexer: John Noble
Picture Researcher: Jenny Veall
Production: Peter Hunt and David Hearn

Typeset in Scala and Glypha LT
Printed and bound in China

CONTENTS

INTRODUCTION

PALE BLUE DOT

On 14 February 1990, Valentine's Day, the space probe Voyager 1 had reached a distance of some six billion kilometres from the Earth, speeding away from us on its epic journey past the planets and into outer space. The spacecraft had just enough of its precious fuel left to undertake one more special manoeuvre, and on that day the mission controllers gave the instruction. The legendary astronomer Carl Sagan had persuaded them to turn Voyager around, to face back towards its distant home for one last time. Travelling at the speed of light, the signal took 6 hours to reach the spacecraft, but when it did, Voyager responded dutifully. As it turned, laid out before its tiny camera – the camera that over a 13-year mission had faithfully captured the most breathtaking and inspiring images of worlds we never knew could be so strange – was the entire Solar System. Very slowly, one by one, Voyager took a last photograph of every planet it could see, and over the succeeding three months transmitted them back to Earth. The result includes one of the most powerful images of all time: the planet Earth, pitifully small, almost indistinguishable from the thousands of points of light of the stars behind it, appears as a tiny dot of pale blue, less than a single pixel wide, caught in a beam of reflected sunlight bouncing off the shiny surface of the spacecraft. It is a humbling and deeply inspiring picture – all humanity, all our achievements, our futures, all our hopes and dreams, captured in this pinpoint of light.

Left:
A Titan-Centaur rocket soars into the Florida skies on 5 September 1977, breaking the sound barrier as it launches the Voyager 1 space probe on its journey to the edge of the solar system.

But the image of the pale blue dot represents the pinnacle of something else that is special: the knowledge that enabled it to be captured, there, then, at that moment in history. Voyager itself is a product of two millennia of scientific achievement. The material chemistry that built its foil covered skeleton; the mastery of energy that thrust it into space atop a controlled explosion of rocket fuel; the mathematics that understood the opportunity to use a unique alignment of the Solar System to accelerate the probe in a slingshot from one planet to the next; and the quantum physics that allowed its electronics to send back its precious observations of new worlds. Also on board the little spacecraft is a very special cargo, placed there just in case, sometime in the far distant future, it should encounter an alien intelligence. It is a special kind of phonographic record, a gold-plated copper disc containing photographs that summarize our hard-won scientific knowledge – including chemical and mathematical definitions, anatomy and geology. For good measure there are scenes of life on Earth and even sounds such as greetings in 55 different languages, a performance of Beethoven's 5th symphony and a blast of Chuck Berry playing "Johnny B Goode".

SCIENCE NOW

Today, Voyager continues to travel, now some 17 billion kilometres from home, pushing at the edge of interstellar space and carrying with it human scientific achievement in microcosm. Its remarkable mission is the product of questions that have been asked by every human since time immemorial – the great questions: who are we; where did we come from; what are we made of; what is out there? The story of how humanity has sought to answer those questions is the story of science.

In telling that story we will unfold how the modern world was built. For science is embedded in our lives so completely that today we barely notice its presence. Our mobile communication networks depend on orbital mechanics that enables satellites to be positioned in the sky, the chemistry of the rocket fuel that launched them, and the materials that make up the plastics and silicon chips of our computers, phones and batteries. Modern medicine relies not only on intimate knowledge of the biochemistry of each of our cells, but also on an equally deep level of understanding of the atomic structure of matter, in order to scan our organs and bones, to diagnose disease. Our access to the energy that fuels our busy lives depends on our understanding of the geology of the inner Earth and the laws of thermodynamics. Our ability to farm our land and feed our people depends on biologists' capacity to manipulate the process of evolution in the animals and plants that live alongside us. Nothing that we do today is untouched by science, and if we can understand better how that came to be, then we will be better equipped to respond to an uncertain future.

"The history of science is often told as a series of great breakthroughs, revolutions and moments of genius from scientific heroes. But there is always a before, an after and a historical context."

The history of science is often told as a series of great breakthroughs, revolutions and moments of genius from scientific heroes. But in reality there is always a before, an after and a historical context. For science does not happen in a vacuum; it is not set apart in an ivory tower. Science has always been a part of the world within which it is practised, and that world is subject to all the usual complexities of politics, personality, power, passion and profit. So as this story unfolds we will meet characters who worked within the limits of the political and religious climates they knew and were subject to the same pressures as the people who lived alongside them. Only by understanding their world can we understand why the extraordinary advances of science took place when and where they did.

Left above:
The cover of Voyager 1's "golden record" offers instructions on how to build a machine capable of playing it. The disc contains sounds and images of distant humanity, alongside a map to locate the probe's origin.

Left:
A grainy image of Earth from more than six billion kilometres away represents one of the pinnacles of human achievement, as well as a salutary reminder of our true place in the cosmos.

RIGHT PLACE, RIGHT TIME

What is often seen in the history of science is that discoveries emerge from different people at or around the same time. Charles Darwin developed his theory of evolution by natural selection over a number of years in the mid 1800s. Meanwhile, another man, Alfred Russel Wallace, quite independently developed a theory that was in many ways remarkably similar. Why? Well, the idea that evolution was something that might explain the diversity of the natural world was already being much talked about; both Darwin and Wallace were part of a world eager for travel and exploration, and both had seen things that puzzled them on their voyages; both had read a book by Thomas Malthus that explained how populations are kept in check by famine and disease. But above all, both were part of a historical climate, a society that was driven by overt competition. Victorian life was gripped by the concept of progress, and throughout all layers of society could be felt the consequences of success or of failure to adapt to the rapidly changing commercial and industrial environment. It was in the context of this combination of factors that each of them was inspired to conclude that the driving force behind evolution might be the pressure of natural selection.

It is not just historical events that provide the framework within which scientific understanding has advanced. Technological inventions and discoveries have been critical to the story of science, both directly and indirectly. In the early 15th century the invention of the printing press (more accurately printing with moveable type), attributed to Johann Gutenberg in Germany, resulted in a cascade of consequences for science. The effects of this single event rippled out across the known world, and on over centuries of time, launching the first information revolution. Before the printing press, knowledge was effectively rationed because of the high costs of making books, which had to be copied by hand. At the start of the 1400s, an educated person might have owned a handful of books, if that. After the invention of printing, it was possible to have a library; a collection of books on different subjects – books that did not necessarily all agree with each other. Printed books were used to carry the latest thinking on all subjects – scientific, literary and religious – encouraging the idea of questioning traditional authority. But there is another aspect to the invention of the printing press that can be easily overlooked. Book reading was now a private activity; it did not have to take place in church and it did not have to be supervised. It was one of the many changes that helped to create the more individual, questioning minds that came to make scientific achievements.

More directly, the availability of new technology has frequently resulted in leaps forward in areas of science in which it was suddenly possible to measure and observe things that hitherto would have been unthinkable. The most obvious examples are the invention of the telescope and the microscope, which transformed the understanding of the cosmos at one end of the scale, and the workings of the living cell at the other. Yet often technological invention, and the scientific advances that followed, emerged for very unscientific reasons, such as in the case of the steam engine, which came about as a response to commercial demand – it was the work of practical engineers trying simply to make some money. But once the steam engine was in existence, it became an object of study in itself, as scientists tried to understand the principles of power and energy that enabled it to work. The result of this was the discovery of the fundamental laws of physics that underpin the nature of the Universe.

Above:
The success of tobacco helped to fuel the worldwide search for new species of plants to exploit. The knowledge uncovered changed our view of the origins of life itself.

FOLLOW THE MONEY

As in most walks of life, financial pressures have played a significant part in shaping the progress of science. The story of Galileo's use of the telescope to study the heavens was largely driven by money. When he first heard rumours of the miraculous new invention, the spyglass, the reason he leapt so enthusiastically into action was his difficult financial situation – he was at the time a middle-aged professor of mathematics with limited prospects and badly needed to improve his status and finances. News of the invention of the telescope must have seemed almost heaven sent, an opportunity to impress a new patron amongst the wealthy families of 17th century Italy. He was, of course, utterly unaware of how his brilliant use of the device would come to change the science of the Cosmos.

On a rather grander scale, when explorers and collectors set off on botanical expeditions into the unknown during the course of the 17th and 18th centuries, at least part of their motivation was to find new species of plant that they could exploit. Early adventurers had shown the fabulous wealth that could be obtained from the discovery and sale to the Old World of plants like tobacco. This search for botanical gold led to the unfolding of new knowledge about life across the planet, and fuelled a new understanding of where we, as animals, came from.

"The availability of new technology has frequently resulted in leaps forward in areas of science in which it was suddenly possible to measure and observe things that hitherto would have been unthinkable."

CHARACTER COUNTS

All scientific discoveries owe much to the particular historical context that their discoverers find themselves in. Some, however, also seem to depend critically on the discoverer's character. A good example of this is Johannes Kepler. At the beginning of the 17th century, working alone in Prague, Kepler discovered three laws of planetary motion that would in time transform our view of the Solar System. He would not have done what he did without the political and religious changes brought by the Reformation, which undermined belief in established authority and drove him from his home. He needed the financial and political support of the Holy Roman Emperor, Rudolph II – an Emperor obsessed with horoscopes who had the money to pay for stargazers. And he certainly needed data that had been painstakingly collected over many years by his colleague Tycho Brahe. But he also needed a spark of genius and an obsessive personality.

At a time when almost everyone believed the Earth was at the centre of the Universe, Kepler believed passionately that the Sun was the symbol of God and produced a force that drove the planets around itself. To prove this, he realized that he had to describe orbits for the planets that would match what was seen in the night sky. This is where his obsessive nature proved to be so important, because this was complex and unbelievably tedious work. He kept at it for five long years, producing hundreds of pages of calculations before he finally reached a solution. As he later wrote, "If thou dear reader are bored with this wearisome calculation, take pity on me who had to go through it seventy times."

Leonardo da Vinci was a very different sort of character. He too was fascinated by the stars, but he was also fascinated by a great many other things. This, in a way, was his problem, for he was rarely ready or willing to focus long enough to complete what he had started. Along with his painting and his inventions, he produced some of the finest anatomical drawings of the human body ever made, with the intention of creating a textbook of anatomy. The problem was that, like so much else he did, he never finished it. By the time his work was rediscovered, others had done what he set out to do, and he remains forever described as a man who was ahead of his time.

Others who have left their mark on science have shown a much more ruthless streak than either Kepler or da Vinci. Sir Isaac Newton, a strange and obsessive character, took his reputation so seriously that he was prepared to destroy that of anyone who crossed him. Robert Hooke, his intellectual rival for many years, is just one of those who endured his anger and vindictiveness. Newton's malevolent behaviour, however, pales in comparison to the competitive drive of the inventor Thomas Edison. Wanting to demonstrate the superiority of his method of electrical power distribution, Edison quietly supported the development of the electric chair – which was reliant on a competing method – to show that the alternative was lethal, and thus to discredit the work of his rival.

SHEER CHANCE

Scientific advances are the product of the same social pressures and subject to the same human triumphs and failings that beset all other walks of life. But blind chance has also played a significant role throughout the history of science. As Louis Pasteur, the father of germ theory and codeveloper of pasteurization, once famously said, "chance favours the prepared mind". The point is that it is important to be in the right place at the right time but that chance alone is not enough. One of the most famous examples of a serendipitous discovery was that made by Alexander Fleming.

In 1928, Fleming went on holiday, leaving behind a petri dish with bacteria growing on it. Somehow fungal spores being investigated by one of his colleagues in another part of the building made their way into his laboratory and settled on the dish. The weather, it so happened, was balmy and created perfect conditions for both bacteria and fungus, so that when Fleming returned from holiday and picked up the dish he was able to say to himself, "Now I wonder what is killing the bacteria?" Luck was certainly on his side; he was extremely fortunate to discover the antibacterial properties of penicillin. But the point is that he had spent a large number of years looking for something that would kill bacteria, so he swiftly recognized the significance of what he saw. Instead of throwing the dish away, he investigated.

"Each explanation that science has offered of how the world works is very largely a product of its time."

The chemist William Perkin also made a discovery that was utterly unexpected, and demonstrated the flexibility and wit to exploit an opportunity when it came along. In 1856, aged just 18, he was in his parents' house trying to produce a synthetic version of the anti-malarial drug quinine. Instead of a quinine substitute, he found that he had accidentally created a chemical that could dye a bit of cloth an intense purple. Many chemists might have put this discovery aside as being of minor importance, but Perkin realized its commercial potential. Not only did he make his fortune, he also created a new and important branch of industrial chemistry.

As we shall see as the story of science unfolds in this book, over centuries the forces of history, personality, money and technology have all come to bear on the moments in which each of the great scientific advances has occurred. Each explanation that science has offered of how the world works is very largely a product of its time.

Right:
Louis Pasteur's lifesaving research into killing microbes began with his early experiments in the production of a disease-free beer. His work exemplifies his belief that "chance favours the prepared mind".

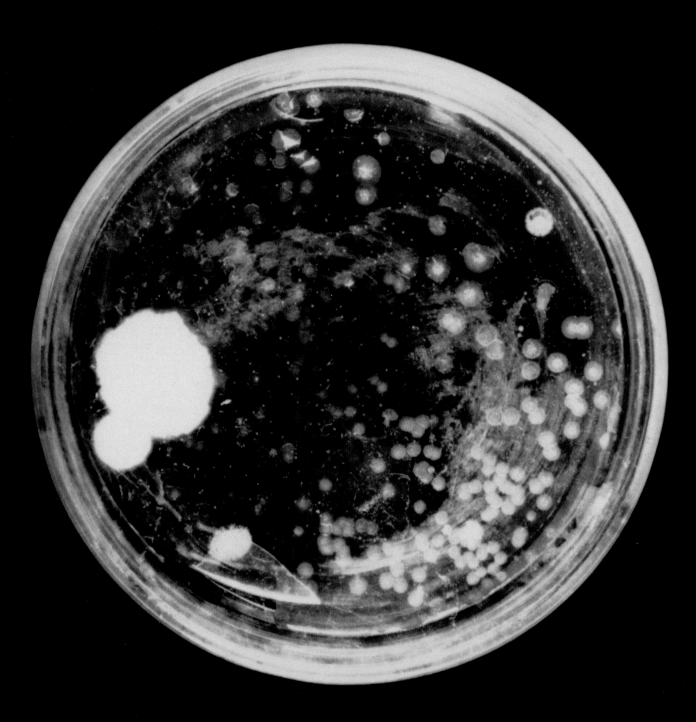

Print of the culture plate which started
the work on Penicillin

UNCOMMON SENSE

But as science has advanced it has also often struggled to find acceptance among the wider population. Part of the problem is that we are used to having opinions about the world and, on the whole, having those opinions respected. Yet when it comes to science, the opinions of the many do not count. Science is not, in that sense, democratic. Many of its truths are counterintuitive, and our everyday experience of the world is largely irrelevant to its progress. When it comes to science, common sense will only take you so far. Can there possibly be an attraction between the Earth and the Sun that operates instantly and across huge tracts of empty space? How plausible is it that whole countries sit on top of great slabs of rock that slowly float across the surface of a molten core? Can we, and every creature on Earth, really be descended from the same single cell? The world around us seems reassuringly solid, so what does it mean when scientists tell us that everything is made of atoms and that atoms themselves consist almost entirely of emptiness?

In daily life, we end up taking these things largely on trust, as for the most part they do not affect us directly. But when there is conflict between evidence and our self interest or our personal observations of the world, problems can arise. Today, the evidence in support of manmade climate change is overwhelming, yet throughout the debate over global warming many people have remained unconvinced that it is a real phenomenon – partly because the science is so complex and the detail so uncertain, and partly because the consequences for our lives are so difficult to accept. Similarly, Darwinian evolution is the most robust of scientific theories, but many people find it easier and more reassuring to believe that something so intricately constructed as life on Earth must have had a designer. It is hard for us to accept that everything around us, from the beauty of the butterfly to the complexity of the human eye, could have evolved simply through chance and the pressure to survive. But the scientific evidence that evolution by natural selection has occurred is unassailable.

The scientific method, which is today practised in all corners of the globe, builds explanations based on evidence, and when new evidence emerges that does not fit with the model, the explanation must change. That is how science moves on. Ultimately, however doubtful the motives, however influential the economic, political and personal forces that have born down on the men and women who practise it, and however unpalatable the conclusions are, the development of science has depended on following where evidence leads. The path that we have followed in search of answers to our big questions has been an epic one, full of rich characters and inspirational ideas, moving in fits and starts, pursuing blind alleys, surmounting great hurdles and frequently taking unexpected turns. It has been one of the most colourful and exciting stories that it is possible to tell.

Left:
A photograph of Alexander Fleming's original culture of the Penicillium fungus. Though serendipity played a great part in the initial discovery, it was the methodical work of other scientists that turned it into a viable treatment.

Cosmos

WHAT'S OUT THERE?

If, on a clear night, you are lucky enough to find a secluded place from where you can gaze up at the night sky and be awed by the enormity of what you see, it can be a humbling and moving experience: a canopy of stars spread above you, inspiring wonder; everything appearing quiet, still, and peaceful. Yet we are actually on a large rock spinning around its own axis at over a thousand kilometres an hour. We and our companion, the Moon, are also travelling in a giant loop around the Sun at about a hundred thousand kilometres an hour and we are held in this endless swirl by a warping of space-time that we call gravity. We are just part of one tiny galaxy that, along with thousands upon thousands of others, is being pulled through space towards a gravitational anomaly in the intergalactic void. These are not intuitively obvious concepts to grasp; indeed they go against all our common-sense experience, because to all intents and purposes we live on a static surface that is daily lit by a rising and setting sun. It is therefore not surprising that it took much time, much toil, and much anguish to arrive at the understanding of the cosmos we have today, and our answer to the question: what's out there? It is a remarkable story and it centres on a time of terrible upheaval some four hundred years ago.

Left: This Hubble Space Telescope image captures details within a barred spiral galaxy similar to our own Milky Way. This independent system containing billions of stars is so far away that its light has taken 70 million years to reach Earth.

THE COURT OF THE EMPEROR

The forces of change that swept across 16th-century Europe were powerful, fast moving, and deadly. The Protestant Reformation, started in 1517 as a protest against corruption in the Church of Rome, brought more than a century of turmoil to the fledgling nations of Europe. Yet through all the violence that ensued one deeply reassuring idea held sway, as it had for millennia: the Earth was, self-evidently, at the centre of everything. Around our secure and fixed world moved the heavens, with their pinpoints of starlight, their wandering planets, and the life and light-bringing Sun and Moon; and all of this had been arranged there for us by God, whether that god was mediated by the pronouncements of the Pope or interpreted literally from a Lutheran Bible. It was an idea in which the power and authority of both Church and State was vested. Yet quietly, throughout the turmoil, building gradually as the century drew to a close, that view was to be revolutionized, and the certainty and authority that it brought was to be shattered.

The roots of that transformation can be uncovered among the characters of the flamboyant court of the Holy Roman Emperor Rudolf II, King of Bohemia, King of Hungary, Archduke of Austria and Moravia, who reigned from 1576 to 1612. He was a melancholic and withdrawn ruler whose primary interests lay in the occult and in learning, rather than in the machinations of politics and government. But above all, Rudolf was a collector. He gathered a menagerie of unusual animals, original artworks, clocks, scientific instruments, and a botanical garden, and had a special wing built at Prague Castle to house his spectacular collection of "curiosities", which included almost everything from a gem-encrusted rhino horn to what he thought were the feathers of a phoenix, a mummified dragon, and the horn of a unicorn.

Rudolf also collected people. His passion for learning attracted some of the best minds of the age to his court, and his tolerance of different religious beliefs created an atmosphere of dialogue and discussion around him. It was to his court in Prague that travelled two of the most significant people in the story of our understanding of the heavens – and one of the oddest couples in the history of science. One was an arrogant Danish aristocrat with the finest assembly of astronomical instruments of the age, and the other was a poor, German mathematician on the run from religious turmoil.

Below: The Basilica of St. Vitus towers over an enormous maze of buildings that makes up Prague Castle.

THE IRASCIBLE DANE

Tycho Brahe
1546–1601

His false nose and apparently enlarged left eye can be seen in the portrait.

The year 1560 was historically unremarkable. But on 23 August there occurred an event that made a deep impression on the first half of our odd couple – a total eclipse of the Sun. In Copenhagen it was seen only as a partial eclipse, but the fact that the event had been predicted on the basis of tables of observations of the movement of the stars and Moon seemed of great significance to a young Danish aristocrat boy, Tycho Brahe, and stimulated a passion in him for star gazing. In his mid-teens he studied law, but also began to buy up astronomical instruments and books, beginning a lifetime of making observations of the night sky. What struck the young Tycho was the variability of the observations recorded in star tables that had been passed down from the ancients. Aged just 17, he had written: "What is needed is a long term project with the aim of mapping the heavens conducted from a single location over a period of several years." That was what he set out to do, and by the end of his life he had provided the fundamental evidence that was needed to answer the question of what is out there.

Tycho moved on through several of the great universities of Europe – Wittenberg, Rostock, Basle, and Augsburg – studying astrology, alchemy, and medicine, and amassing an extraordinary collection of astronomical instruments. As a student, however, an event occurred that literally marked him out for life. Tycho got into a fierce argument with another scholar while at a dance. We do not know for certain what the argument was about, but it ended in a duel, fought in the dark, the result of which was that he lost a portion of the bridge of his nose. For the rest of his life he wore a special cover made of metal – it is said he wore a copper alloy for everyday use and gold or silver as formal wear.

"By the end of his life he had provided the fundamental evidence that was needed to answer the question of what is out there."

Right: One of Tycho's most famous achievements was an enormous model of the celestial sphere, 180cm (6ft) in diameter. Sadly, it was destroyed in a fire at Copenhagen University in 1728.

The silver nose underlined Tycho's reputation as a high-handed and irascible aristocrat, but his reputation as an observer of the heavens was not established until a few years later with the appearance of a remarkable new star in the night sky. By then he was back in Denmark, and on the evening of 11 November 1572 he was walking back to his house from his alchemical laboratory when he noticed a very bright new object in the constellation Cassiopeia. Tycho had spotted what we now know as a supernova – the explosive death of an ageing star. Indeed, it was he who coined the term "nova" when he published a book on the discovery, *De Nova Stella*. As he observed it assiduously over the following months SN1572, as it is known today, gradually dimmed, but what he saw was quietly shocking for the model of the Universe that he and all around him held to at the time. As he built up a dossier of observations it became apparent that it was indeed a star, not a planet, nor a comet. Here was something new appearing in what everyone believed was the "crystal sphere" of the fixed stars that had been created, perfectly formed, by God. For a new star suddenly to emerge in the heavens was a chilling idea.

Below: Newborn stars illuminate huge regions of pinkish star-forming gas in this Hubble Space Telescope image of an "irregular" galaxy.

SPHERICAL HARMONY

The earliest of the ancient civilizations all shared the same fundamental view of the cosmos: that the Earth lay at the centre. The Sumerians, the Babylonians, and the Egyptians all had the Sun, Moon, stars, and planets revolving around us. The specific explanations varied from ancient society to ancient society, but the one that came to dominate the minds of Europeans was established by successive generations of Greek philosophers. We tend to lump all these thinkers together in one group of "Ancient Greeks", but in reality they were spread across many centuries, and their theories of the cosmos were refined over a period of more than six hundred years.

The first known idea of the stars being fixed to a sphere, or hemisphere, rotating around us is attributed to Anaximenes of Miletus in the 6th century BC. At that time his model saw the Earth as a kind of flat disc, or flat-topped cylinder that floated like a cork on the air. Pythagoras of Samos – the same Pythagoras whose theory we use today to calculate the area of a triangle – changed the disc to a globe, and placed it at the centre of concentric spheres, one for the Sun, the Moon, and each of the planets, with the "fixed" stars at the furthest distance. For Pythagoras, the distances separating the spheres were of great importance, and he saw the seven planetary spheres, (including the Moon and Sun) and the sphere of the stars being separated in the same seven ratios as those of the musical scale. It was this notion that gave us the concept of the "harmony of the spheres" that was to resonate for two millennia.

The model that later became fixed stemmed from a proposition laid down by the philosopher and mathematician Plato in around 400 BC. For Plato, the circle was a perfect form, and he was convinced that the planets, the Sun, and the Moon must therefore revolve about a spherical

"Eudoxus required 27 concentric spheres to explain the movements in the heavens, but that was then refined by his contemporary, the great philosopher Aristotle, into a model of greater perfection."

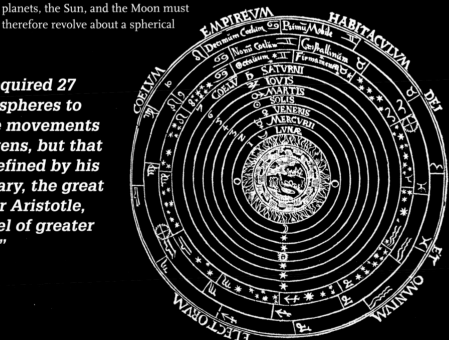

Earth in circular orbits. His students were left with the challenge of creating a model to explain his philosophy, and one Eudoxus offered an ingenious solution of multiple concentric spheres. The orbit of the Moon illustrates the idea. To explain its apparent motion through the heavens, the Moon needed three spheres: one rotating every day, to explain the rising and setting; a second rotating every month, to explain the movement through the zodiac (its movement against the stars); and a third rotating every month on a slightly different axis, to explain its variation in latitude. The problem that was obvious to the Ancient astronomers was that the planets behaved in a strange fashion, sometimes seeming closer, sometimes further away from the Earth, sometimes speeding up, and sometimes slowing down or even appearing to go backwards. The word "planet" comes from the Ancient Greek for "wander". Eudoxus required 27 concentric spheres to explain the movements in the heavens, but that was then refined by his contemporary, the great philosopher Aristotle, into a model of greater perfection. In an attempt to make sense of what was observed, he placed 55 concentric spheres around the Earth, each responsible for some specific movement of the heavenly bodies, always in the perfect eternal motion of a circle, as they passed through the substance out there that he called the "aether". And at the outside he placed the "Unmoved Mover", the force that centuries later came to represent the all-powerful Christian God.

All that, and the bulk of this chapter of this book, could have been rendered irrelevant had the ideas of Aristarchus, also of Samos, caught on some two hundred years later, at the height of Athenian domination of Greece. Essentially, he had it all worked out. He placed the Sun at the centre of the cosmos, with the Earth and other planets circling it, in the same order as we know them today, and with the fixed stars at a far greater distance away. But his idea did not stand up to the withering logic of the time. For a start, he was unpicking Plato and Aristotle, which was not the thing to do, but the real reason Aristarchus's idea did not catch on was simply that it seemed so self-evidently wrong. If the Earth were moving through space, why would an object thrown upwards, or fired like an arrow, come straight back down, instead of landing at a

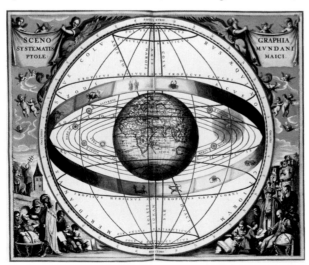

distance away as the ground you were standing on whizzed away from it? Or, if nothing else, if the Earth were moving through space, why could we not feel a howling wind, or even a breeze, as we do on a ship moving at sea? No. Common sense told everyone that Aristotle had the right explanation.

Left: An illustration of the Ptolemaic view of the Universe, with Earth at the centre, the Sun and planets moving around it, and the entire system enclosed in a sphere of stars including the ring of the zodiac.

CERTAINTY

This all represented more than the musings of philosophers and mathematicians. The works of Aristotle became one of the philosophical pillars passed down into Christian teaching in the centuries that followed, through the era of the Roman Empire and its subsequent collapse. Kept alive by the schools of Greek philosophers known as the Peripatetics and the Neoplatonists, Aristotle's thinking greatly influenced both Arab scholars in the near East and the early figures of Christianity, such as Saint Augustine, whose teachings lay at the heart of Church rule. Aristotelian cosmology, or the Christian interpretation of it, took hold because it made sense, and the Unmoved Mover of the celestial spheres clearly was a representation of the all-powerful God. So it was that Aristotle's description of the heavens became an unchallengeable feature of Church authority for the next thousand years.

With the disintegration of the Roman Empire and its relocation to Constantinople, much of Ancient Greek knowledge was lost to what we now think of as Europe, but it became scattered piecemeal through the Middle East as traders moved in and out of that great city, which became the heart of the Byzantine world. It was avidly studied by the Arab scholars, who gradually spread across North Africa and Spain following the emergence of Islam in the 6th century AD. Gathering knowledge was part of the process of attaining spiritual perfection for Islamic scholars, and the emergence of large libraries – storehouses of both old and new learning – marked out the long rule of the Muslim caliphs, as a demonstration of their striving for completeness. Texts from both Greek and Latin were translated into Arabic, studied and built on. Arab science bequeathed us new forms of mathematics, an understanding of optics, and medical practice, and it also ensured that the ancient observations of the cosmos were maintained.

> "Arab science bequeathed us new forms of mathematics, an understanding of optics and medical practice, and it also ensured that the ancient observations of the cosmos were maintained."

Above: Classical knowledge frequently returned to Europe in the form of beautiful illuminated Arabic manuscripts.

CIRCLES WITHIN CIRCLES

The extent of the Arabic store of knowledge became apparent with the fall of the city of Toledo in Spain in 1085, a turning point in the long conflict between Christians and Muslims that was playing out across the Mediterranean. When the city fell to the Christians, vast libraries were discovered holding a multitude of texts, including Arabic translations of Ancient Greek works long forgotten in Christian Europe. One of these was a remarkable book known as the *Almagest*. It had been compiled in the 2nd century AD by a mathematician called Claudius Ptolemy, who made an extraordinary attempt to reconcile the cold reality of observed planetary and star movement with the circles and spheres of Aristotelian philosophy.

PTOLEMY'S EARTH-CENTRIC UNIVERSE

Claudius Ptolemy's *Almagest* contains a set of star tables that could be used to calculate the past and future positions of the planets in the sky, and a catalogue of the 48 constellations visible to the Ancient Greeks. The system of the Universe that he worked out sometime after 100 AD had the Earth at the centre, then the Moon, then Mercury, Venus, the Sun, Mars, Jupiter, Saturn, and lastly the sphere of "fixed stars". For them all to be orbiting a stationary Earth required the planets to be turning in small circles (epicycles) around points in space that themselves revolved in larger circles (deferents) around the Earth. In refining their earliest models, the Greeks had introduced more of these epicycles, including some that were offset from their centre of rotation (eccentrics). Now, to try to fit the observed planetary movement, Ptolemy added the notion of points in space (equants), from which an orbit would appear to be circular. It was amazingly complicated.

Right: A late medieval illustration shows how Ptolemy's system of epicycles and deferents explains the apparent movement of the planets.

Ptolemy lived in Alexandria, in Roman Egypt, and drew on his predecessors' observations of the stars to put together his book of astronomical tables and a description of his model of the Universe. If Aristotle's 55 spheres were a philosophical dream, then Ptolemy's solution was a mathematical nightmare, which included many epicycles (wheels within wheels) in order to make the observed movements of the planets fit the common-sense view that the Earth was stationary at their centre. But it was this complexity that was passed down through the generations that followed. Copies of Ptolemy's *Almagest* began to spread among the astronomical community of Western Europe, and this combination of Aristotelian philosophy and Ptolemaic cosmology was the accepted wisdom within which Tycho's observations had begun to play out.

Tycho's reputation as one of the leading astronomical figures of the day had been established by the new star of 1572. The Danish king, Frederick II, who feared losing the prestige that such a prominent figure would bring to Denmark, offered him an extraordinary reward: his very own island, a bare little spot just 5km (3 miles) long, called Hven, which lies between Denmark and Sweden, and whose 370 inhabitants even today reach it only by a long ferry ride from the Swedish mainland. On it, he built what was then the most spectacular observatory in the world, Uraniborg (the Castle of Urania), and it was there, in 1577, that he made yet another cosmologically disturbing observation – that of a comet. Once again, his careful plotting of its movement enabled him to say for certain that its path was one that passed right through the "crystal spheres" that were thought to hold each of the planets in orbit around the Earth. It was yet another suggestion that all was not well in the established model of the heavens.

THE POOR MATHEMATICIAN

Johannes Kepler
1571–1630

Over the years, Tycho became increasingly despotic and arrogant, and he ruled Hven as a private fiefdom, even imprisoning his tenants for misdemeanours. Eventually he fell out with the new king and left Denmark in 1597. After two years travelling through the German states, he and his entourage arrived in Prague, and he was granted an audience with Rudolf II. The result was his installation as Imperial Mathematician to the Holy Roman Emperor in the castle of Benatky, some 35km (22 miles) from Prague, where Tycho set up his instruments and continued his observations. And it was to Benatky that he invited Johannes Kepler – the poor mathematician and the other half of our odd astronomical couple – to be his assistant.

Kepler was the polar opposite of Tycho Brahe. Born in 1571, not far from today's city of Stuttgart in Germany, he was poor, and his entire life had been caught up in constant turmoil. Wars were neither few nor far away in 16th-century Europe, and his father scratched a living as a mercenary, coming and going from the family until Johannes was in his teens, before disappearing forever. Aged only four he caught smallpox, and was left with poor eyesight and terrible skin problems for the rest of his life. He later plotted a genealogical record of his family and his early years, and in places it reads as a miserable catalogue of his ailments and misfortunes. But he also noted childhood experiences of the cosmos, which clearly made an impression on his young mind. As a little boy in 1577, he recalled being taken out at night to stare at the great comet – the very one that Tycho was at the same time tracking, far to the north on Hven.

Scientific biographies can often overstate the academic brilliance of their subject, but in Kepler's case there can be no overstatement – he really did stand out as a gifted mathematician. He was sent to the University at Tübingen,

CASTLE OF THE STARS

The Castle of Urania (the muse of astronomy) on Hven was constructed on four floors, with rooms for Tycho Brahe and his family, other rooms for visiting astronomers to work in, and attic spaces for Tycho's students. To support his work, the site had an alchemical laboratory in the basement, a medicinal herb garden, fish ponds, an arboretum, a tannery, a flour mill, a printing press, and a paper mill. At the heart were the instruments, the most famous of which was a giant quadrant with a 2m (7ft) radius, mounted to the wall, and thought to have framed a portrait of the great astronomer himself at work, accompanied by his sleeping dog.

The key to compiling accurate astronomical tables was regular observation of the movement of the star or planet relative to the position on Earth. Observation in Tycho's day was extraordinarily difficult, carried out as it was with the naked eye, but his quadrants and sextants were the most accurate that existed before the era of the telescope. To maintain the observatory and his lavish lifestyle, Tycho received annually what has been estimated as over one per cent of the entire income of the Danish crown – more, relatively speaking, than NASA receives from the US government today.

where he also studied physics and astronomy, and there he was introduced, on the quiet, to a recently formulated theory of how the motions of the planets, Sun, and Moon could be explained. His professor, Michael Maestlin, publicly taught his astronomy students the established model that came from the Greeks – Ptolemy's description of a Universe with the Earth at the centre of everything, complete with all its epicycles. But unofficially, he also introduced them to the more radical Copernican model, where everything revolved around the Sun. This "heliocentric model" was an idea that Kepler seized on.

THE POLISH CANON

Nicolaus Copernicus's real name was Mikołaj Kopernik, but he Latinized it, following the practice amongst intellectuals of the time. He was a canon at Frombork in Poland, although the post was effectively a sinecure and he spent years at a time away from the town, studying in Italy. Sometime around the turn of the 16th century, Copernicus became aware of the concept of the heliocentric Universe as it had been espoused by the Ancient Greek thinker Aristarchus, possibly through references in Latin and Greek texts that he studied in Padua. It was not a new idea. Apart from its disputed status amongst Greek philosophers, it had also been debated by Arabic and early Indian thinkers. What Copernicus saw in it was a way to make the complex cosmology that had been handed down from the Ancient Greeks a great deal more elegant.

Contrary to what is often claimed, Copernicus was no revolutionary. All he set out to do was to improve on what the ancients had done before him. The difficulty was that, not wishing to depart too far from the Greeks' deeply held philosophy, he still assumed that the planetary orbits were perfect circles (which they are not) and so, to try to fit the data from real observations, his model needed epicycles too. The result was that it ended up almost as complicated as the Greek model he was trying to simplify.

COPERNICUS COMPLEXITY

By adopting the idea that the planets no longer revolved around the Earth, Copernicus not only got rid of the complexity of Greek epicycles, but also, by arranging the planets in what we now know as their correct order from the Sun (Mercury closest, then Venus, Earth, Mars, Jupiter, and finally Saturn – the others had not then been discovered) he made sense of the different lengths of time that each of them took to complete their astronomical cycle. However, his model was not the simple representation of the Solar System that every schoolchild is taught today. The difficulty was that he assumed that the planetary orbits were perfect circles and to try to fit the data from real observations, he arranged the planets, including Earth, as revolving around a single point in space – but this was not the Sun. Instead it was a position just slightly removed from the centre of the Sun – a distance of about three times the diameter of the Sun, to be precise. The net result was that Copernicus ended up with almost as many, if not more, epicycles than the Greek model he was trying to simplify.

He first made his model known in a short paper that was circulated in manuscript form in 1510, the Commentariolus. This caused no great upset in Catholic Church circles, and indeed was favourably discussed at the Vatican after a lecture attended by the Pope. There is an irony in that while the Protestant Reformation had opened a new chapter in the questioning of the authority of the Church, and the world view that it espoused, at the same time the Protestant faiths themselves tended to a more literal reading of the word of the Bible. So it was that by 1543 when Copernicus finally published *De Revolutionibus*, the small book that outlined his theory in all its detail, it was the wrath of the originator of the Reformation movement, Martin Luther, that was feared, not that of the Catholic Church. The publication was handled by a Lutheran minister called Andreas Osiander, and unbeknownst to the ageing Copernicus he decided to play safe. He inserted a preface declaring that the theory contained within the book was only that – a mathematical model that might be helpful in describing the motions of the planets, not a suggestion that this was really how they moved. It was a subtle, but important distinction. A mathematician could offer any model that he liked as a mathematical exercise, but to step into the realm of explaining the world that God created was a dangerous step to take. Copernicus was almost certainly unaware of the last-minute addition to his book. The story goes that as he lay on his death bed, the first copy of the printed book was placed in his hands, and he died at peace.

"To step into the realm of explaining the world that God created was a dangerous step to take."

But the book was read. Sometime in the 1450s, in Mainz in Germany, Johann Gutenberg and his associates invented a process of printing with movable type, one of the most significant inventions in human history. Within a generation it had revolutionized the spread of knowledge and information throughout the nations of Europe. There were presses in all the major cities and the rate of publication of books grew exponentially. Where manuscript books, copied painstakingly by hand, had been cherished, prized, sometimes hallowed objects, held by the very wealthy and powerful, now printed books were available for almost anyone to own. The book market was dominated by printed Bibles, but quickly it spread wider into practical texts and "how to" books. By the end of the 16th century, some 150 million books, on everything from animal husbandry to the zodiac, had been published across Europe. This was the transformed world into which Copernicus's book had emerged, and it soon found its way into the libraries of most of the intellectuals and academics with an interest in understanding the motions of the heavens. By the early 1600s copies of the first edition were to be found as far afield as Sweden, Sicily, Spain, and southern Ireland.

Above: An illustration of the new model of the Universe proposed by Copernicus. The Sun lies at the centre, with Earth as the third planet, orbited by the Moon. Around the outside, the cosmos is enclosed by a shell of fixed stars.

MYSTERIES OF THE UNIVERSE

Kepler's interest in the heliocentric model had as much to do with mysticism as with pure mathematics. He was a brilliant mathematician, but he was also an astrologer. In fact, the two skills went hand in hand at that time, and astrology for many was a very serious business. Political leaders, like Rudolf in Prague, would take few decisions the night and day brought by the Sun, the monthly cycle of the Moon, and the annual seasons all had readily discernible effects on life here on Earth, so why should not the other planets and stars have powerful effects on our lives as well? Viewed from the time, such thinking made perfect sense. Astrology therefore required carefully calculated descriptions of planetary motion, and accurate predictions of their future position. After he graduated at Tübingen, Kepler took up a position as a professor of mathematics in Graz, in Austria. His salary was low, and he supplemented his income by using his mathematical skill to cast horoscopes. Indeed, his long description of his family and upbringing formed part of an extended horoscope that he prepared for himself. In later years he was to look back at aspects of his life and seek astrological explanation for some of his darkest moments, such as the deaths of two of his children, and the unhappiness of his marriage.

With eyesight too poor to observe the stars and with little access to observational data, it has been argued that Kepler inevitably found himself adopting the approach of the Ancient Greek philosophers in simply thinking about how the heavens might be explained, in isolation from any evidence.

"This was an idea that drove him for the rest of his working life, and brought us fundamental laws of planetary motion that we depend on still."

However that may be, he specifically recorded a moment, in July 1595, when a flash of inspiration came to him while drawing on the blackboard in a lecture he was giving on the conjunction of Jupiter and Saturn in the zodiac. It had suddenly occurred to him that the Copernican heliocentric view of the Universe fitted perfectly with the shapes of the five perfect geometric solids that featured in the philosophy of Plato and the Ancient Greeks that followed him. Perhaps, Kepler thought, the Universe was constructed according to fundamental geometric rules. From the 21st-century perspective, it was complete nonsense, but to Kepler it must have seemed like divine revelation; the new Copernican heliocentric model and the geometry of the ancients combined to confirm his own blend of religious belief and cosmology. He was convinced that the Sun at the centre was the expression of God, that the fixed stars were Christ, and that the planets and space between were the Holy Spirit. This was an idea that drove him for the rest of his working life, and in so doing brought us fundamental laws of planetary motion that we depend on still today.

Below: A sequence of photos captures the changing appearance of the Moon over the course of a month.

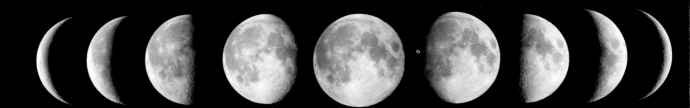

Kepler published this strange mixture of religious belief and astronomy in 1596 in a book called *Mysterium Cosmographicum* (*The Mysteries of the Universe*). It was the first published defence of Copernican theory, and it is a reflection on the fledgling nature of scientific thinking at the time that it had as much to do with mysticism as it did with pure mathematics. Kepler, then in Graz in Austria, began sending copies of his answer to life, the Universe, and everything to prominent astronomers in Europe, and his reputation grew. When a copy reached Tycho Brahe, the two began corresponding, and the upshot was his invitation to become an assistant to the great Dane in Prague. It was a timely offer because Kepler's personal circumstances were dire; he had become virtually a refugee from the violent religious turmoil that was engulfing Europe.

KEPLER'S GEOMETRIC RULES

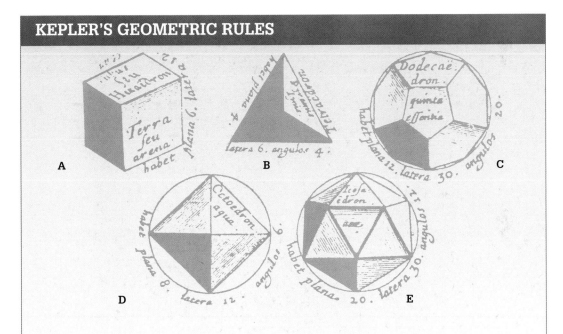

In his 1596 book, *Mysterium Cosmographicum*, Kepler imagined a geometric cube placed between the "spheres" that carried the orbits of Jupiter and Saturn. The relative distance between these orbits (as defined by Copernicus) was such that the points of the cube touched the inside of the Saturn "sphere", while the surface of the Jupiter "sphere" just touched the insides of the cube. A bizarre thought it may have been, but to Kepler it was of enormous significance, for the cube was one of the five "Platonic solids" in the philosophy of the Ancient Greek, Plato. The other four Platonic solids were the tetrahedron (B), octahedron (D), dodecahedron (C), and icosahedron (E). Kepler placed one of each geometric shape between each of the planetary "spheres" (working inwards from Saturn the order was: cube, tetrahedron, dodecahedron, icosahedron, and octahedron) and then put a sphere in the centre for Mercury. With a bit of fudging here and there, and allowing for the poor quality of astronomical observations that were available, he found that he could get the planets to sit roughly in their correct positions.

MEETING OF MINDS

So it was that three timelines of history came together in Prague. The work that was done there was driven together by the politics of Reformation and princely courts; it was fuelled by the technological advances of the printing press and the finest astronomical instruments; and it questioned the intellectual achievements of the ancients that had been handed down across generations. None of the participants knew it, but this particular time and place would mark a key step in changing our understanding of how the Universe works, and of our place in it; it would begin to provide an answer to the question of what is out there.

In February 1600, Kepler came to Prague with his books, including a copy of *De Revolutionibus*, and a determination to make the maths of his geometric universe work perfectly. In Benatky Castle, Tycho was as controlling as ever and only allowed his new assistant access to his star data in little dribs and drabs – nothing like enough to verify a model of the Universe. Then, 18 months after Kepler arrived, Tycho died suddenly. It is said that his bladder was strained to bursting after he refused to leave a banquet to relieve himself. That is an unlikely cause, and in truth no one really knows what killed him, but Kepler's life was transformed, for he found himself appointed in Tycho's place to the position of Imperial Mathematician to the Emperor Rudolf II.

More importantly for the story of science, he now had access to all of Tycho's vast collection of astronomical observations. It was a treasure trove of star data; by the end of his colourful life, Tycho Brahe had mapped accurately the positions of 777 of the "fixed stars". To put that in context, the stars we can see with the naked eye from any point on Earth number no more than about 2,500, with good visibility. Next time you glance up at the night sky, it is worth reflecting on that lifetime achievement of dogged observation. It is often said that as a result of his hours of straining to look at tiny points of light in the sky, one of Tycho's eyes grew bigger than the other. Unlikely as that is, a portrait of him does exist that shows his left eye decidedly larger than the right – although whether that was a real representation, artistic licence, or self-promotion remains unknown.

> *"It was a treasure trove of star data, for by the end of his colourful life Tycho Brahe had mapped accurately the positions of 777 of the 'fixed stars'."*

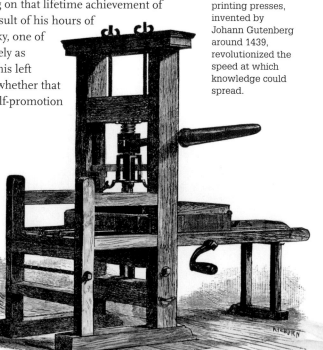

Below: The first printing presses, invented by Johann Gutenberg around 1439, revolutionized the speed at which knowledge could spread.

95 THESES

Martin Luther
1483–1546

The Protestant Reformation had been sparked by Martin Luther, an Augustinian monk, who was also a professor at the University of Wittenberg in Germany. Luther and others in the northern European states had become increasingly vocal critics of what they saw as corrupt practices that pervaded the Catholic Church. In particular they were concerned about the sale of indulgences. An indulgence is the remission of penance for sins that have been freely confessed, but by the 16th century they had become a commodity for sale, and something of a small industry, with travelling "pardoners" distributing printed sheets that offered shortening of penances, time off from purgatory, or even release from hell itself – all for an appropriate price. In the early 16th century Pope Leo X had stepped up indulgence sales as part of a fundraising campaign for the renovation of St Peter's Basilica in Rome, and it was probably this move that prompted Luther to act. In 1517, he famously nailed his 95 "Theses on the Power and Efficacy of Indulgences" to the door of the castle church in Wittenberg. But while the issue of indulgences may have triggered the original protest, the reforms that Luther was calling for were of a fundamental nature, challenging some of the very tenets of the Catholic faith. Inevitably the Church hit back; Luther was declared a heretic at an assembly of the Imperial Estates of the Holy Roman Empire at a small town called Worms, on the Rhine (the "Diet of Worms"). The Vatican denounced him and the stage was set for more than a century of conflict, culminating in the bloody Thirty Years War that saw Protestants and Catholics in open combat throughout Germany and much of the rest of Europe until the end of the 1640s.

For Johannes Kepler, the consequences of the Protestant Reformation were extreme. The religious division across the Holy Roman Empire had pretty much settled as a north–south divide, with Reformation dominating the north and Catholic faith holding sway in the south. But the ruler of each individual state was allowed to impose the religion of his choice. This meant that the approved religion could change at a stroke with a new duke or a prince, and not all were as tolerant of different faiths as was Rudolf in Prague. Kepler, as a staunch Lutheran – albeit holding Copernican views of the cosmos – was caught up in just such a shifting pattern in Catholic-dominated Graz. At first his position was tolerated but a change of ruler in 1596 brought a crackdown on the Protestants and Kepler's life became increasingly uncomfortable.

"The reforms that Luther was calling for were of a fundamental nature, challenging some of the very tenets of the Catholic faith. Inevitably, the Church hit back."

Right: A modern spaceprobe image of Mars. Kepler's efforts to explain the red planet's complex path around the sky helped to uncover the way the Solar System works.

KEPLER'S LAWS

One of the most obvious difficulties with the ancient geocentric model of the Universe was the observed movement of the planet Mars, which, if it was travelling around the Earth, could only be explained by its sometimes speeding up and slowing down and occasionally going back on itself. Confident of his mathematical ability, Kepler thought he would have the orbit plotted in a week. Instead it became a labour of years, but it was to prove to be the key in unlocking the real secret of planetary motion.

Sticking with the ultimate goal of proving his own cosmological explanation of the Universe, Kepler steadily unpicked one after another of the complexities that Copernicus had put in place to make his model work: the orbit offset from the centre of the Sun, the uniform speed of the planet, and then, finally, the fundamental shape of the circular orbit. It was this last shift in thinking that proved to be key. At first he had tried an egg-shaped orbit, but that didn't fit the observations. But when he replaced the egg shape with an ellipse, everything fell magically into place. At a stroke, all the epicycles and retrograde loops vanished in a perfect mathematical description of the orbit of Mars around the Sun. Kepler concluded that all planets move in elliptical orbits, with the Sun at one focus. This became known as Kepler's first law of planetary motion. Two further laws also emerged from his mathematical mind, defining the relative speed at which the planets move, and relating the time that they take to orbit the Sun to their distance away from it. All three laws were to prove crucial to the work of Isaac Newton, some 75 years later, in defining the forces that govern the Solar System as we understand it today.

Following page: In 1604, Kepler (like his mentor Brahe a generation earlier) observed a "new star" in the constellation of Ophiuchus, the serpent-bearer. Marked with a letter "N", it lies in the figure's foot on this illustration from Kepler's treatise on the star. Today, it is known to have been a supernova – the violent death of a star far more massive than the Sun.

Below right:
Kepler's
Rudolphine Tables
allowed
astronomers and
astrologers to
calculate the
positions of the
planets relative to
the background
stars with
unprecedented
accuracy.

For Kepler, however, there was a nagging disappointment: the elliptical orbits of the planets, although clearly the answer that fitted the observations, meant that his own cosmology, the great scheme of nested Platonic solids, did not easily fit the facts and he struggled on for the rest of his working life, trying to reconcile the two ideas. In essence, his book, *Astronomia Nova*, in 1609 swept away the remnants of the ancient, orthodox, geocentric model of the Universe. It should have caused a revolution. But it did not. Few people read it, and those who did found it largely impenetrable, for the mathematics was wrapped up in his mystical thinking, and not easy to follow.

Kepler devoted the rest of his life to the task he had been charged with on Tycho Brahe's death: to compile all of the Dane's star data in the form of a book of astronomical tables. He completed them, and in 1627 published the *Rudolphine Tables*, a compendium that eclipsed Ptolemy's *Almagest* and became the best star charts available for several generations. Despite the Earth-changing nature of his mathematical work, at the time of his death in 1630, these tables were seen to be his most important legacy. His revolutionary answer to the question of what's out there passed almost unnoticed. Getting people to accept such a radically different answer to the question would require a showman, a fighter, someone with a massive ego who was a great deal more persuasive and better connected than Johannes Kepler.

KEPLER'S LAWS OF PLANETARY MOTION

1. (A) Planets move in elliptical orbits with the sun at one focus of the ellipse (not the centre).
2. (B) The orbital speed of a planet varies so that a line joining the sun and the planet will sweep over equal areas in equal time intervals.
3. (C) The amount of time a planet takes to orbit once around the Sun is called the period, P, and it's related to the orbit's size, A. The Period, P, squared is proportional to the semi-major axis, a, cubed. This says that the further away a planet is from the Sun the longer the time it will take to go around it once.

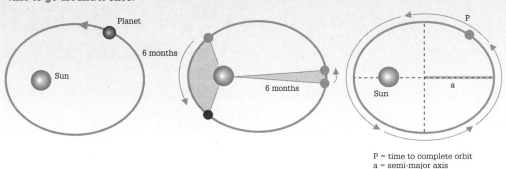

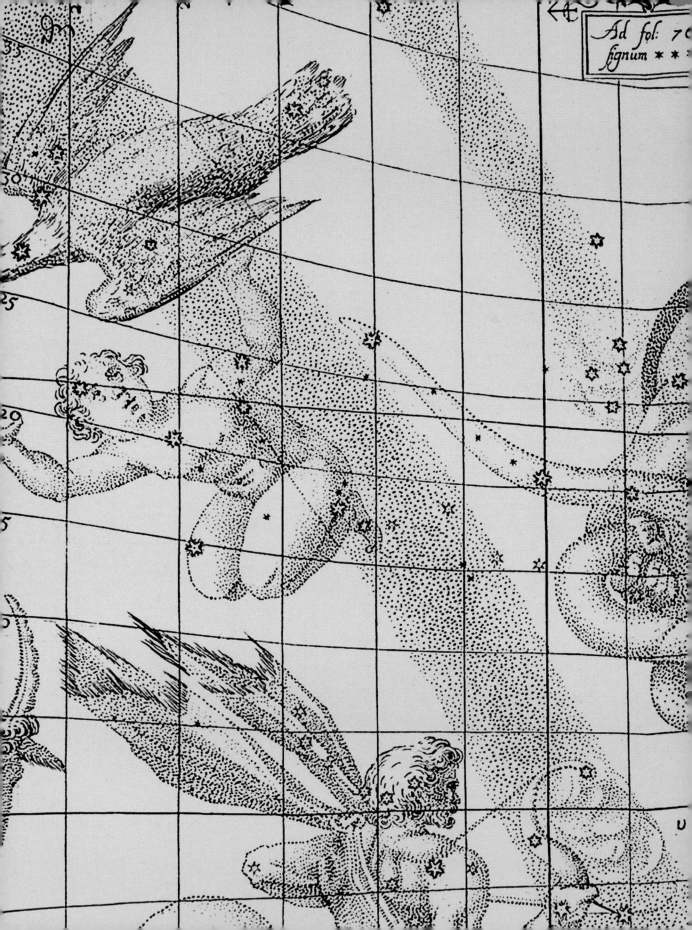

DUTCH GLASSES

War continued to be a backdrop to science throughout the 16th and 17th centuries. Late in 1608, there was a peace conference in the Netherlands, which ushered in what became known as the "Twelve Years Truce" in the Eighty Years War, in which the Protestant Dutch fought to shake off the rule of Spain. At the conference, Prince Maurice of Nassau was shown a "spyglass" by Hans Lippershey, a spectacle maker from Middelburg. Spectacles had been around for hundreds of years by then, and spectacle makers had mastered the tricks of shaping concave lenses to correct short-sightedness and convex ones to assist reading. What Lippershey did was to put one of each together to produce a rudimentary telescope. That same year he had tried to get a patent on his invention, but it was refused because two other claimants came forward at roughly the same time. Why the idea suddenly emerged then in Holland is unclear, but its potential on the battlefield was immediately obvious, and the presence of many diplomats at the demonstration meant that news of it spread rapidly across Europe as emissaries travelled home with the glad tidings of peace. By early 1609, small spyglasses could be bought at the Pont Neuf in Paris, and by the summer they had reached Italy.

"The device was very simple: a tube with a convex lens at one end and a concave one at the other, producing a magnification of about three times."

The device was very simple: a tube with a convex lens at one end and a concave one at the other, producing a magnification of about three times. When news of the spyglass reached a professor of mathematics in Padua, he thought he could probably make one too. The mathematician was Galileo Galilei, a combative, bombastic man determined to succeed and ever on the lookout for a chance to improve his position amongst people who mattered. Here was an opportunity to impress his potential patrons, and so he planned to build a spyglass for the Doge of Venice, under whose authority he lived and worked in Padua. In the summer of 1609 came news that a Dutchman was about to present a telescope to the Doge, and in an extraordinary piece of practical craftsmanship Galileo had his own version built in 24 hours and then improved it to eight times magnification. This he managed to present to senators of the republic, ahead of his competitor, in August of that year. The result was a sensation. Galileo wrote that several times he and Venetian gentlemen of influence climbed the highest bell towers in the city to view distant objects greatly magnified. His submission to the Doge emphasized the advantage it would give the maritime state in its constant need for defence against attack from the Ottoman Turks, "allowing us at sea to discover at a much greater distance than usual the hulls and sails of the enemy, so that for two hours and more we can detect him before he detects us". This was how the world worked in the time of Galileo: a significant gift was offered to a patron, together with a gentle hint at the need for personal gain. The Doge responded generously, offering to double Galileo's salary and confirm his position in Padua for life. However, the small print was less attractive: life meant exactly that; he would not be allowed to move, and the salary would never change. So Galileo set his sights elsewhere.

Left: Galileo Galilei 1564–1642. This portrait dates from 1636, during the period of Galileo's house imprisonment.

Right: A pair of Galileo's early telescopes are now displayed side by side at the Museum of the History of Science in Florence.

PERFECT VISION

Venice at the turn of the 16th century had passed the peak of its Renaissance glory. The great Venetian rowing galleys, which had dominated Mediterranean trade and numbered some 3,000 at their peak, had been overtaken by the sea-going ships that now plied the oceans to the west. The fall of the Byzantine city of Constantinople to the Ottomans had resulted in a reawakening of classical knowledge as both people and texts crossed the Aegean to safety in Italy, bringing understanding that helped to fuel the flowering of the cultural and humanist explosion of the Renaissance. At the same time the need to find other routes to the Far East, to bypass the Ottoman power, opened up the drive for exploration and the discovery of the New World – the Americas. The result was that economic power was shifting away from Italy towards northern Europe. But Venice was still a cultural powerhouse, and among its crafts was glass. As far back as the 1200s, Venetians had brought sodium ash back from the Middle East and this was one of the key ingredients of the remarkably pure crystal glass that their glassmakers produced on the island of Murano in the Venice lagoon. Indeed, the Murano craftsmen were prohibited from leaving Venice, in order to try to keep their special skills secret. It was this glass that helped Galileo to make dramatic improvements to his telescope over the autumn of 1609; by the end of the year he had achieved a magnification of 20 times.

Galileo had also begun turning the telescope to the heavens, starting with a study of the Moon. What he saw, and revealed in a series of eight drawings across its different phases, was startling. Instead of the perfect smooth sphere defined by Aristotle's cosmology, he saw rugged mountains and jagged edges to its circumference and shadow-line. He went on to look at other heavenly bodies. In January of 1610, he saw what seemed to be three small stars in a line across Jupiter; three nights later one had disappeared. Three nights after that it was back, in a different position, and he also

Right: Spaceprobes have revealed the four tiny points of light that Galileo discovered orbiting Jupiter as worlds in their own right. Here the volcanic moon Io (left) and icy Europa hang in front of Jupiter's own turbulent cloudscape.

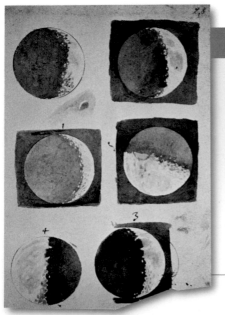

GALILEO'S MOON OBSERVATIONS

These pioneering observations were first made using the telescopes that Galileo had built in 1609. Galileo saw that the terminator (the line between the Moon's night and day sides) was sometimes irregular (top) and sometimes smooth (bottom). He deduced that the irregularities were due to mountains on the Moon, the first time Earth-like objects had been discovered in the heavens. This challenged the existing world view that said the heavens were perfect and unchanging. This page is from a 1653 edition of *Sidereus Nuncius* (March 1610).

saw a fourth. He realized that these "stars" must be orbiting Jupiter. He had discovered the planet's moons. Galileo was already favourably disposed to the Copernican view of the cosmos, but this was powerful evidence, albeit indirect, in its support. For if Jupiter had moons that orbited around it, then the Earth's special position at the centre of everything was no longer so special. Galileo decided to publish his discoveries, and wrote them up in a short, very accessible book written in Italian rather than Latin and including pictures. He called it the *Siderius Nuncius* – the "Starry Messenger". Ever the opportunist, he dedicated it to the Grand Duke Cosimo II di Medici, and proposed naming the four new satellites of Jupiter after the four Medici brothers. As a tactic, it worked. By the summer of 1610 he had been appointed as court mathematician and philosopher to the Medicis in Florence, at a hugely increased salary. His future status and wealth were secured.

"If Jupiter had moons that orbited around it, then the Earth's special position at the centre of everything was no longer so special."

THE TRIAL

The story of Galileo's later life – his clash with the Church, his trial for heresy, and his imprisonment – has stood out as a turning point in the history of science, but the reality is not quite as simple as it is often portrayed. In the end, it was less about heresy than about authority. For what Galileo began to do was to stray into territory that the Church believed should be firmly under its control. The observations Galileo made with his telescope gave him more confidence in Copernican theory. Not only did he see the moons of Jupiter and the uneven lunar surface, but also he observed the complete phases of Venus – its changing illumination as it orbits the Sun, falling into and out of shadow, as it passes in front of the Earth. His findings, at first, were accepted for what they were, simply observations with no consequent interpretation, and Galileo had a successful audience with the Pope in 1611, with much glory being attached to his patrons in Florence as a result. But thereafter things gradually began to go wrong. He antagonized an influential Jesuit astronomer by over-claiming the discovery of sunspots, and he became more public in his support for Copernicanism. In a letter to the Grand Duchess of Tuscany, in response to her concerned questioning as to whether the theory went against biblical teaching, Galileo famously wrote: "One must not begin with the authority of scriptural passages, but with sensory experience, and necessary demonstrations." This is one of the earliest expressions of what we now think of as the scientific approach – evidence first, deductions follow. But it also reveals that Galileo was beginning to say that his world of scientific study might have equal status to revelation. This was dangerous ground. In 1616, after a Papal commission concluded that Copernican theory was indeed heretical, Galileo was instructed not to "hold to or defend" the heliocentric view.

Left: According to a popular tale, at the end of his trial a humiliated Galileo forced to recant his beliefs, still muttered under his breath "nevertheless, it [the Earth] moves!" Sadly, the story first appears a century after the trial itself, and there is no evidence that it is true.

Above: Galileo published his *Dialogue* in Italian rather than Latin, ensuring his ideas would reach the largest possible audience.

There it lay, but once again Galileo indulged in a public spat with Jesuit astronomers, in a critique he wrote of their theories on comets. And in a book that followed he underlined his philosophy that the laws of the Universe "are written in the language of mathematics". Yet again, an audience with the new Pope Urban VIII, who had been a personal friend, went well, and Galileo was encouraged to write a book that compared the two world views – the Ptolemaic and the Copernican – providing, of course, that he did not claim that the Copernican model was true. The result in 1632 was the *Dialogue Concerning the Two Chief World Systems*. In it he set out the arguments for and against each model in the form of a debate between three characters. It is generally agreed that he went too far, allowing himself to be identified with the defender of Copernicus, and putting the case for Ptolemy into the mouth of a character he called "Simplicio" – the simpleton. The Inquisition saw its chance and the trial of Galileo, for breaking the injunction not to defend Copernicanism, was engineered. Famously, he confessed guilt and recanted; the *Dialogue* was banned. Galileo was declared a heretic and placed under house arrest for the remainder of his life. But most importantly, Vatican authority over knowledge was restated: truth lay in the teaching of the Church.

FATHER OF SCIENCE

As is ever the way, the long-term outcome of suppression was not what the factions inside the Church might have hoped for. The book was smuggled out of Italy and the intellectual community across Europe read it avidly. Within a generation, few educated people still believed that the Sun travelled round the Earth, although nothing that Galileo had done could be regarded as incontrovertible proof. Indeed, that had to wait for another century, when instruments were good enough to observe stellar parallax – the shift in the relative position of the stars and the Earth, as we orbit the Sun. But after Galileo, and after Kepler, the question became less to do with how the planets move as with what holds them in place.

Arguably, Galileo's greatest work was in an entirely different realm of "natural philosophy". He had spent his life doing mathematical and experimental work in the realm of physics, fluid science, and mechanics. He laid out mathematical laws of the movement of projectiles and the acceleration of falling bodies; he famously proved that objects of different sizes and weights would fall at the same speed (although stories that tell of him dropping cannon balls and feathers off the Leaning Tower of Pisa are nothing more than stories); and he demonstrated that it was the density of an object rather than its shape that determined whether it would float.

Galileo died in January 1642 at the age of 77. The family villa where he was incarcerated, at Arcetri in the hills above Florence, still carries an air of calm isolation and solitude. For his last decade the old man, gradually growing blind and more infirm, spent his time there pulling together the results of his lifetime's work. Above all, everything he had done was based on observation and experiment – gathering evidence and building arguments from what can be seen and reproduced is at the heart of a modern scientist's work. That is perhaps why Galileo has been called the "father of science".

THE HIGH SEAS

The opening of the sea routes to the Americas by Christopher Columbus in 1492, and the heroic ocean voyages that followed, transformed the world. The 16th century was the era of global circumnavigation, with the first expedition around the world in 1519 heralding an unprecedented territorial expansion of the maritime European countries. The wealth in gold and silver that was extracted from the New World and found its way back to Spain, Portugal, the Netherlands, and Britain was vast. Silver from the Americas alone accounted for a fifth of Spain's national budget in the late 16th century. This new global trade and ocean exploration brought with it a far greater urgency to solve a problem that had been recognized for centuries: how to arrive at an accurate measurement of longitude, a position east or west around the globe. Throughout the early age of expansion there was real economic pressure to find a solution in the measurement of stars, Moon and planets. The Spanish and the Dutch governments both offered prizes for a solution; the French founded the Académie Royale des Sciences, charged with improving navigation, and in England the Royal Observatory was founded in 1675 at Greenwich. Astronomy became the most important scientific endeavour of the time, and the topic of everyday conversation amongst the intelligentsia of Europe.

Left: When Columbus's fleet landed in the Bahamas in 1492, he initially believed he had arrived in the Far East. The opening up of East-West trade routes in the following century only added to the need for an accurate method of determining longitude.

LATITUDE AND LONGITUDE

Early astronomical instruments, such as the cross-staff, the astrolabe, or the quadrant, were easily sufficient for determining a sailor's latitude (position north or south of the equator), by measuring the angle between the horizon and the Sun or known stars. Longitude is very different. Because the planet rotates once every 24 hours, the solution ultimately lay in the ability to know the time at your home port when it was, say, noon at your position on the ocean. The time difference will tell you how far round the globe you are. But a reliable, accurate, ship-board clock (the chronometer, famously perfected by a Yorkshire carpenter, John Harrison) was only available from the late 1700s, and then barely affordable until the mid 19th century. An alternative method was known as "lunar distance", and depended on knowing how the position of the Moon would appear relative to the stars at different points around the globe. This required the compilation of lunar distance tables, based on thousands of observations. Until recently ships always carried lunar distance tables with them, in case of the failure of their chronometers.

THE BET

By the late 16th century, global trade had brought great wealth to the City of London. The repeated attacks of plague had passed, and the Great Fire of 1666 had triggered a huge programme of building and renewal. Everywhere in the capital, coffee houses had sprung up and people gathered in them to discuss the matters of the day; to do deals, to argue and debate rival theories of the world, to listen to stories of distant lands, and above all to exchange information. In 1684, three men regularly met at Jonathan's and Garraway's coffee houses and it may well have been at one of these two establishments that they agreed a bet. Christopher Wren, architect of the rebuilding of London, Edmond Halley, yet to become linked with the comet that today bears his name, and Robert Hooke, the curator of experiments at England's national academy of science, the Royal Society in London, were debating the question of what actually held the planets in their elliptical orbits around the Sun. At that time the favourite explanation was magnetism. During the conversation they challenged each other, for the sum of two pounds – or about 250 cups of coffee – to demonstrate that the force, whatever it was, obeyed the inverse square law, namely that the force gets weaker in proportion to the distance, squared, of the planet from the Sun. This was an idea that had been discussed for a number of years, but mathematically, it was a very difficult thing to prove.

"They debated what held the planets in their orbits. The favourite explanation was magnetism."

Below: Coffee houses were popular meeting places in the 17th and 18th centuries, and the scene of many scientific innovations.

GRAVITY

Later that year, Halley visited Cambridge and called to see Isaac Newton, the Lucasian Professor of Mathematics at the University, and a man with a formidable scientific reputation, especially for his work on optics. Halley put the question of the inverse square law to Newton and was astonished when Newton told him that he had already proved it – but he just could not find the paperwork there and then. The encounter prompted Newton, encouraged by Halley, to go back over his work of some 20 years previous and compile it into a book. By rights, the publication should have been funded by the Royal Society, of which both Halley and Newton were influential members, but the Society had recently lost substantial sums on publishing a natural history of fish, by one Francis Willoughby, that had failed to sell. So Halley stepped in and paid for the printing himself (and was repaid in copies of the fish book). It is fortunate that he did so, for Newton's work, published in 1687, stands as one of the greatest contributions to science of all time. It was called *Philosophiae Naturalis Principia Mathematica* – the *Principia* for short.

The history of science is often told as stories of individual brilliant minds, flashes of inspiration, and men leaping out of baths shouting "eureka". The reality is very different: ideas emerge into the zeitgeist and become talked about; technological advances make things possible to see or understand; historical events open up opportunities or pressures for change. There is always a context. Isaac Newton and the advances he made were part of all that, but he is truly one of the few greats to whom the word "genius" can perhaps be applied. Born in 1643 at Woolsthorpe in Lincolnshire, Newton was a complex, insular man. His childhood and youth had been miserable. He once even threatened to burn down his house with his mother and stepfather inside. He was educated away from home, and then brought back at the age of 16, when his stepfather died, to run the farm. He was a terrible farmer and so was sent to Cambridge University as a "subsizar", a position where he had to act as a servant to other students to earn his education. Then, to escape the threat of plague, he returned to Lincolnshire in 1665, and it was there, over just two years, that he did the science and mathematics for which he would become most famous. He went back to Cambridge to become a fellow of Trinity College in 1667 and matured into a secretive, reclusive, and obsessive man – but an intellectual giant. It is always easy to think of Newton and his achievements as a shining example of scientific thinking at work – a beacon of the Enlightenment and so on – yet he had unusual religious beliefs that made him a near heretic and saw his principal aim as to explain the mind of God. What's more, the bulk of his life's study was spent in pursuit of alchemy. In reality, he sits uneasily between the occult and mysticism on one side and the beginnings of true science on the other.

By the time Edmond Halley visited him in 1684, Newton had already shown that the colours of the rainbow could be passed through a prism to recombine as white light, thus proving the nature of the spectrum. He had already invented a new form of mathematics, which we now know as calculus, which enabled the analysis of movement. And he had, as he told Halley, already worked out the nature of the force

Isaac Newton
1643–1727

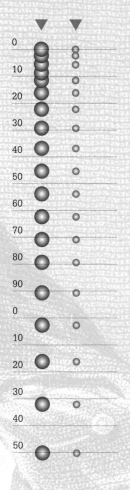

Above: The law of falling bodies, famously identified by Galileo, demonstrates that, air resistance aside, two bodies of unequal weight will accelerate downwards at the same rate in Earth's gravitational field, and hit the ground after the same time.

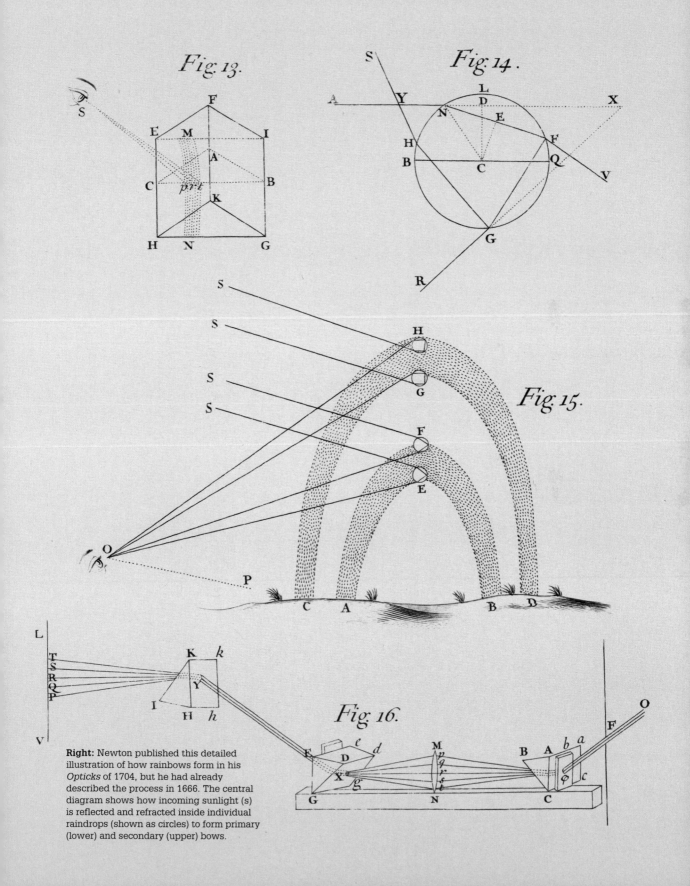

Fig. 13.

Fig. 14.

Fig. 15.

Fig. 16.

Right: Newton published this detailed illustration of how rainbows form in his *Opticks* of 1704, but he had already described the process in 1666. The central diagram shows how incoming sunlight (s) is reflected and refracted inside individual raindrops (shown as circles) to form primary (lower) and secondary (upper) bows.

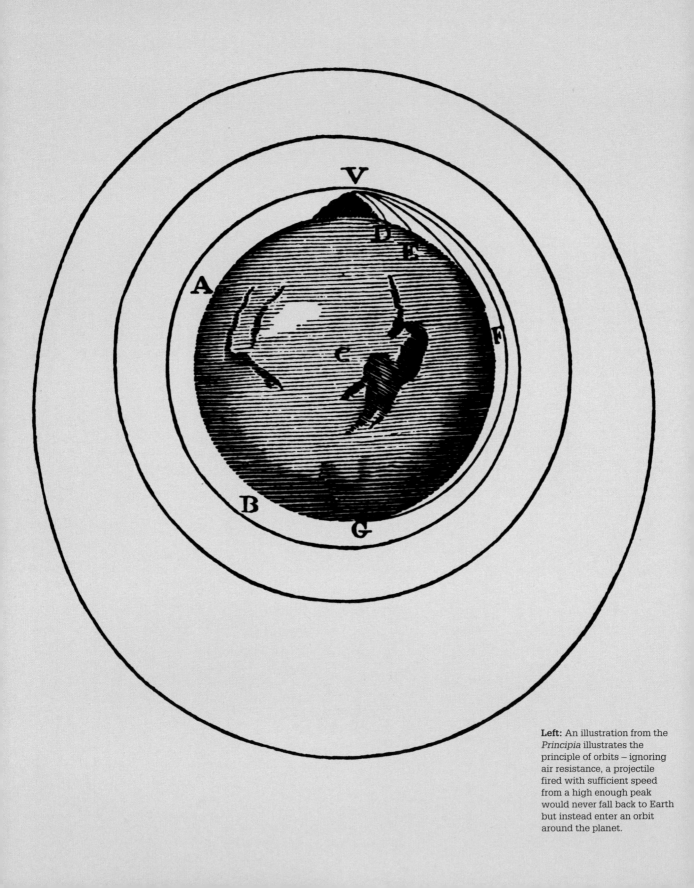

Left: An illustration from the *Principia* illustrates the principle of orbits – ignoring air resistance, a projectile fired with sufficient speed from a high enough peak would never fall back to Earth but instead enter an orbit around the planet.

that holds the planets in orbit around the Sun. He had done the fiendishly complicated maths to show that the force would indeed result in elliptical orbits, fitting in with Kepler's laws of planetary motion. And he had combined this mathematical description of the orbits with the results of Galileo's experiments on the movement of projectiles and falling bodies, to explain why the planets stay in orbit.

UNIVERSAL LAWS

The *Principia* defined the nature of the force of gravity and set out laws of motion for everything from a cannon ball, to the Moon, to a planet orbiting the Sun – or to an apple falling from a tree. The famous story of his watching the apple fall as a young man, and realizing the same force could be extended to the Moon, is almost certainly a complete fiction – a story made up by Newton himself in later life to ensure that he retained credit for the discovery, and to immortalize himself. For there was a side of Newton's character that was deeply unpleasant. He engaged in lifelong and vindictive disputes with the first Astronomer Royal, John Flamsteed, and with Robert Hooke of the Royal Society, who himself came close to or even matched Newton's achievements in optics and defining gravity. Newton even carried the dispute beyond the grave. After Hooke's death, Newton, by then president of the Royal Society, allowed Hooke's portrait to be "mislaid" during a move of premises, so that the image of his arch-rival was lost to posterity.

The most powerful thing about Newton's laws is their universal nature. For Newton, his work brought him closer to his goal of seeing the mind of God, but for the history of science it marks a rare turning point. He had shown that the laws of physics can be applied to everything. Today, our modern world depends on our understanding of those laws, for example, in the acceleration of the rockets that launch communications satellites into space and the geostatidnary orbits such

PHILOSOPHIÆ
NATURALIS
PRINCIPIA
MATHEMATICA·

Above: Isaac Newton's *Principia* was finally published in three volumes in 1687. Only around 300 copies were printed in the first edition and about 1,000 in the second.

NEWTON'S LAWS OF MOTION

FIRST LAW   An object will stay at rest or continue in motion unless another force acts upon it. For instance, a ball will keep rolling unless friction slows it down.

SECOND LAW The force acting on an object is defined as its mass multiplied by its acceleration,
$$F=ma.$$

THIRD LAW   For every action there is an equal and opposite reaction. For example, the recoil of a gun as a bullet speeds away.

To show how the force of gravity would hold a planet in orbit, Isaac Newton devised a thought experiment. He imagined a cannon on top of a high mountain, far above the atmosphere. He thought: if a ball is fired from the cannon slowly then gravity will pull it down to Earth. However, if the ball is fired with enough force it will escape gravity and disappear off into space. But if the velocity is just right the ball will keep travelling right around the Earth, held by gravity in a giant orbit, just like the Moon.

satellites adopt around the Earth, enabling our dependence on modern communications to continue. In the century that followed the publication of Newton's work, the idea of universal laws took off, and became applied in many other fields: economics, the living body, even the mind. The word "Newtonian" itself came to represent the notion of perfectly ordered and mathematically defined theories, those worthy of being taken seriously.

FOLLOW THE MONEY

The mid 18th century was a glorious time for astronomy. In pursuit of the solution to the longitude problem, money was poured into the study of the stars. The Royal Observatory flourished, and George III – better known today as "Mad King George" – was an eager supporter of science. Telescopes got bigger, and so did the Solar System. King George may have been bitter at the loss of the Colonies in the American War of Independence, but in 1781 there was some recompense when the astronomer William Herschel discovered a new planet, which he named "George's Star" in honour of his patron. The name was not an international hit, however, and by the mid 1800s the world had settled on "Uranus", suggested by a German astronomer, after the father of Cronus (Saturn's equivalent) in Greek mythology. By then, yet another planet had been spotted – Neptune, in 1846 – whose position had been predicted mathematically in a triumphant confirmation of Newton's laws, because of a perceived distortion in the orbit of Uranus due to the gravitational pull of the unseen planet. The 20th century brought yet another discovery, with Pluto completing the set as the ninth planet in 1930. Sadly for Pluto it has recently been downgraded to a "dwarf planet", so the 21st-century Solar System numbers just eight planets.

Each of these great discoveries was dependent on bigger and better telescopes. It meant that astronomy remained a very expensive science and, as so often throughout this story, the astronomers "followed the money". By the turn of the 20th century, the money was in the USA, which was well on the way to becoming the dominant economic power that it is now. The industrial and financial barons of late 19th-century America, with their enormous wealth, were generous when it came to supporting science. In California, the astronomer George Ellery Hale succeeded in raising funds from the Carnegie Institute to establish a huge observatory on Mount Wilson, high in the San Gabriel Mountains above Los Angeles, and to install a 1.5m (5ft) telescope at its heart – the largest telescope in the world. Its "first light" was in 1908, but before it was even active, Hale

Right: The 100-inch Hooker Telescope was responsible for many important breakthroughs, including proof of galaxies beyond the Milky Way.

Left: Uranus (far left) and Neptune – a pair of worlds that are the farthest planets in the Solar System, discovered thanks to advances in telescope technology and mathematical astronomy respectively.

"At the time
the Hooker
Telescope
came into
action it was
still a matter of
debate as to
whether or not
the Universe
extended
beyond the
Milky Way."

was busy raising funds for an even bigger one. The mountings, housing, and 2.5m (100in) mirror of the telescope were hauled piece by piece up the mountain road by mules, and finally set to work late in 1917. It was known as the Hooker Telescope, after the industrialist benefactor who paid for the giant mirror, and it too held the title for the largest telescope in the world, which it retained until after World War II.

At the time the Hooker Telescope came into action it was still a matter of debate as to whether or not the Universe extended beyond the Milky Way (what we now know as our galaxy). Hale had hired a young astronomer called Edwin Hubble to work on the "100-inch" telescope, and in 1923 Hubble observed a special kind of star, called a Cepheid variable, whose luminosity varies in a precisely known way such that its brightness or dimness can be used to provide a measure of how far away it is. The Cepheid he observed was in the nebula of Andromeda, and Hubble calculated that it was far too far away to be part of the Milky Way. Andromeda, and other nebulae he observed, must be galaxies in their own right. Suddenly, the Universe was even more vast than had been believed – we could actually see countless other galaxies around us.

BIG BANG

Hubble did not stop there. Astronomers by now no longer relied on direct visual observation – the Cepheids had been spotted on photographs taken through the telescope, which allowed for very dim objects to be seen over very long exposure times. By 1929, the 100-inch telescope was hooked up to a spectrograph, capturing light at different wavelengths, and it was this combination that allowed Hubble to make another extraordinary observation. He saw that the light from certain galaxies was very slightly redder than from others. It is a phenomenon known as "red shift", where light coming from an object that is moving away from the Earth will travel at a slightly longer wavelength. The longer wavelength shifts the colour more towards the red end of the spectrum (the spectrum that Isaac Newton had so convincingly demonstrated as forming the constituents of white light). It was clear evidence that parts of the Universe were rushing away from us. Hubble also observed that the further away the galaxies were, the faster they were moving away. We were in a rapidly expanding Universe. He was not the first to suggest that this might be the case, but here was tangible evidence that supported the theory of the "Big Bang" origin of the Universe, which had been proposed only a couple of years before. These historic observations by Hubble are what placed him in the astronomical hall of fame, and meant that his name was given to what is the most powerful telescope of all time, the Hubble Space Telescope, whose images from the furthest reaches of the visible Universe continue to amaze us all and show us, quite literally, what is out there.

Below: The Hubble Space Telescope was launched in 1990 to continue Hubble's investigations of the large-scale structure and history of the Universe.

EINSTEIN

In 1905, Albert Einstein published his theory of special relativity, which defined the speed of light as being constant and began the concept of space-time – it also gave us the famous equation $E=mc^2$, which shows that vast amounts of energy are contained within the tiniest mass (See Chapter Four), an understanding that is at the heart of atomic energy. In 1916, he published his general theory of relativity, which tied together space-time and gravity. He showed that space-time is warped by the presence of matter, and that the warping results in gravity, which makes matter move. In extending the theory to the Universe as a whole, he could only make it fit the prevailing view that the Universe was static by adding in a "cosmological constant" to his equations. Later, when it became clear that the Universe is in fact expanding, Einstein called his introduction of the constant his "biggest mistake" and threw it out. In subsequent years, many experimental measurements have backed up Einstein's theories, and they now lie at the heart of our description of how the cosmos works.

UNCERTAINTY

It is worth reflecting that the Hubble Space Telescope is placed in orbit only through our ability to get practical use out of Galileo's understanding of the mechanics of projectiles, Kepler's laws of planetary motion, and Newton's definition of gravity, but what it and other devices for measuring the make-up of the cosmos have uncovered is not quite as straightforward as those laws once suggested. Einstein's theory of general relativity, put forward in 1916, demonstrated that time and space and gravity can be, and are being, warped throughout the cosmos. And as if that wasn't bad enough, today's cosmologists readily accept the existence of black holes from which no light can emerge; they see the need for 11 dimensions of space-time; they speak of string theory, of membrane theory, of parallel universes, of dark matter, and dark energy, something that no one knows the nature of, but which must nonetheless exist. Yet for all of these seemingly weird phenomena, the mathematics is strong and convincing, however fanciful the ideas may sound.

"Our view of the night sky is no longer of a closed world, but of an infinite, and expanding cosmos of which we are only a very small part."

To the person in the street, modern cosmology offers answers to the question of what is out there that must seem as disturbing and unlikely as the idea that the Earth travels round the Sun must have seemed over five hundred years ago. What has changed is that today science tries to respond objectively to the evidence that it finds, even if that evidence goes against the received wisdom. Our view of the night sky is no longer of a closed world, but of an infinite, and expanding, cosmos of which we are only a very small part. But what is perhaps more important is that our whole outlook has changed. We have gone from seeing with a closed mind, to seeing with one that is constantly being forced to rethink – to make sense of what we find out there, in this terrifying but dazzling universe.

Left: In 1924, Edwin Hubble showed that the beautiful Andromeda Nebula was an independent star system more than 2 million light years away. Today we know that it is a spiral galaxy similar to our own Milky Way.

Right: The Hubble Space Telescope has delivered stunning images such as this portrait of the Crab Nebula.

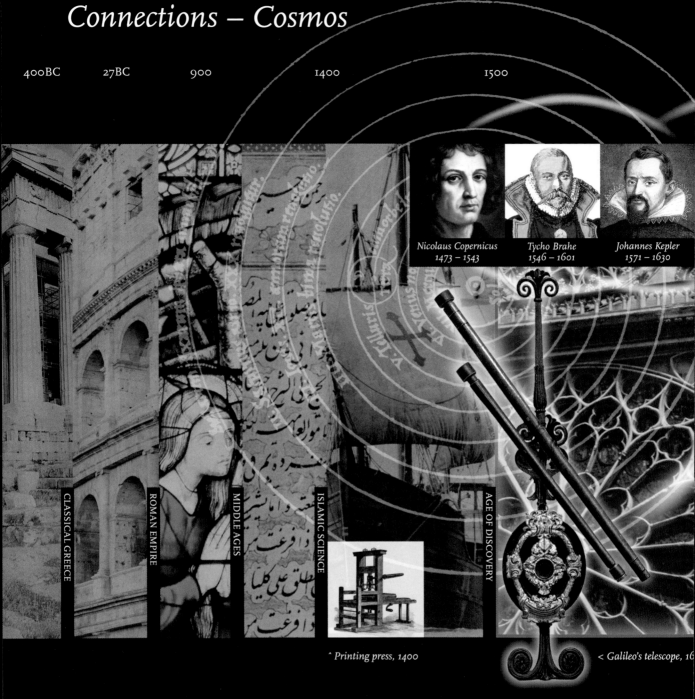

400BC   27BC   900   1400   1500

Nicolaus Copernicus
1473 – 1543

Tycho Brahe
1546 – 1601

Johannes Kepler
1571 – 1630

CLASSICAL GREECE

ROMAN EMPIRE

MIDDLE AGES

ISLAMIC SCIENCE

AGE OF DISCOVERY

^ Printing press, 1400

< Galileo's telescope, 16

Our journey to discover what is out there has been shaped by powerful forces and beliefs. The ancient Greek view of a universe of perfect circles around the Earth was one of the most enduring ideas in human history. Dislodging it depended on accurate evidence gathered by inspired and ambitious men. Using a variety of instruments, including the most precise quadrants available, Tycho Brahe was able to compile incredibly detailed astronomical data over his lifetime. Johannes Kepler used this to prove mathematically Copernicus' earlier theory of a cosmos centred on the sun. But it was only when Galileo refined the newly invented telescope and began to observe the heavens that the new view became widely accepted.

The historical context was crucial too. The courts of the Renaissance encouraged new knowledge and paid for the study of the heavens. The religious upheavals of the Reformation, with two churches competing for control, created

Galileo Galilei
1564 – 1642

Isaac Newton
1643 – 1727

Edwin Hubble
1889 – 1953

Albert Einstein
1879 – 1955

20TH CENTURY

REFORMATION

AGE OF ENLIGHTENMENT

21ST CENTURY

^ Tycho Brahe mural quadrant, 1600

^ Hubble Space Telescope, 1990
^ Mount Wilson Telescope, 1892

an intellectual climate in which it became possible to question authority. The printing press allowed the new knowledge to spread at an unprecedented rate.

After Galileo, the search for ways to look ever deeper into the cosmos continued, as did the work of interpreting what was discovered, by brilliant men such as Isaac Newton. By the early 20th century, telescopes had reached a vast size. The biggest was the Hooker Telescope in the United States, where an astronomer named Edwin Hubble revealed that the universe was expanding. His name has been given to the most powerful telescope of all – the Hubble Space Telescope, whose images have revolutionized our view of what is out there.

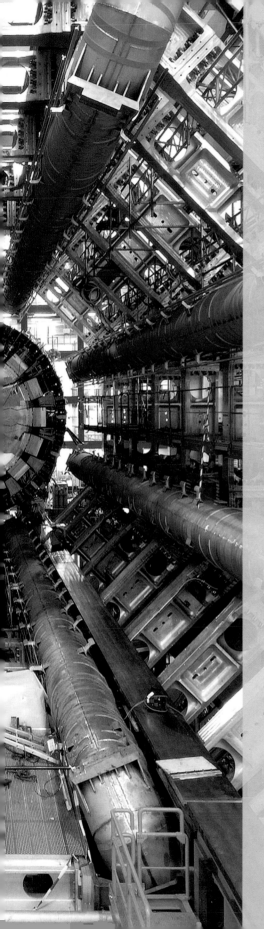

Matter

WHAT IS THE WORLD MADE OF?

The search to understand what the world is made of may seem, on the surface, to be a rather esoteric quest. Yet it has turned out to have had the greatest of practical consequences. Plastics, fertilizers, cars, computers, the internet, genetic engineering, the mobile phone, GPS – all have emerged, at least in part, from attempts to answer this question.

At the heart of it all lies our growing awareness of the way in which all substances, whether solid, liquid or gas, are ultimately composed of countless tiny particles – either molecules composed of even smaller units called atoms, joined together by chemical bonds to form simple or complex compounds, or lone atoms themselves.

Interactions between atoms and molecules forms the basis of all chemistry, but scientists had barely got to grips with this when they were confronted with evidence for another level of deep structure – a subatomic world that, paradoxically, both complicates and simplifies our models of matter. Instead of dozens of elements, most of the material world can now be understood in terms of just a handful of fundamental particles, but the way these few particles behave can be very strange indeed.

Nowhere is our quest to understand this subatomic domain better demonstrated than at the Large Hadron Collider, the world's largest science project, in Switzerland. Here, collisions between streams of particles travelling close to the speed of light aim to recreate conditions last seen in the Big Bang itself, releasing particles that may have stayed locked away since the creation of the Universe.

Left: A view inside the Large Hadron Collider, the world's largest and most ambitious particle accelerator, where streams of subatomic particles are collided with each other at speeds close to that of light in order to probe the deep structure of matter.

ATOMS

Today, it's common knowledge that the world, and everything in it, is made up of atoms – tiny, durable pieces of matter forged in the heart of stars. Atoms are very small indeed. A million of them stacked alongside each other would barely stretch across the letter "a" in this book. Each one consists of a nucleus containing protons and neutrons, around which swirls a cloud of electrons. The exchange and sharing of electrons between atoms in order to achieve stability is the basis of all chemistry, but reshaping atomic nuclei – either by breaking them apart (fission) or forcing them together (fusion) is the realm of atomic physics.

What's more, it's now clear that protons and neutrons are in turn made of even tinier particles called quarks, held together by the exchange of "messenger" particles called gluons. Who knows what further levels of structure still await discovery?

The secrets of the atomic and subatomic world remained hidden from us for centuries, by the limitations of our instruments and our thinking, and by the immense energies required to split successively smaller particles apart. So how did we learn what we now know, and how was the world transformed along the way?

"Atoms are very small indeed – a million of them stretched back to back would be about the size of the letter 'a' in this book."

INSIDE THE ATOM

The world is made of atoms of different sizes. They are measured in picometres, a unit equivalent to a trillionth (1/1,000,000,000,000) of a metre. A group of atoms can bind together to form a molecule, like water (H_2O), which consists of one atom of oxygen bound to two hydrogen atoms. Though they are often thought of as being rather like fuzzy tennis balls, atoms do not have an outside surface as such. Instead they consist of a tiny, dense, positively charged nucleus surrounded by a cloud of negatively charged electrons. Like the inside of a tennis ball, however, the atom is mainly empty space, which means that everything in the Universe, including us, is made mostly of nothing. Despite containing 99.9 per cent of the atom's mass, the nucleus, which is made up of protons and neutrons, is only a minute part of the whole atom – the scale is roughly that of a grain of sand to a cathedral. The number of protons in an atom determines its atomic number and therefore which element it is. A hydrogen atom, for example, contains one proton and so has an atomic number of one. Electrons, on the other hand, determine the chemical properties of an element. A helium atom has two electrons in its "shell", meaning the shell is full, and making it extremely stable and unreactive.

THE LAUGHING PHILOSOPHER

Democritus
c460–370 BC

Let us start with the Greeks and in particular the Laughing Philosopher, "the Mocker", Democritus. Democritus was born in Thrace in around 460 BC. As a young man, Democritus inherited a great deal of money from his father and spent it on travelling and broadening his experience of the world. He went to Asia, the Middle East, and possibly to Egypt. Somewhere along the way he picked up a love and knowledge of mathematics and also a thoroughly materialistic and deterministic view of the world. Democritus believed that everything could be explained by natural laws, if only he knew what these were.

Legend has it that Democritus was sitting at home when he smelt a loaf of fresh bread that a servant was carrying up the stairs. He started wondering how it was that he could smell the bread and concluded that minute particles of bread must escape into the air. He went on to describe a thought experiment with cheese. Imagine, he said, that you cut a bit of cheese in half, then in half again, and so on and so on. Finally you will arrive at something you cannot cut, which Democritus called "atoms". The table is hard, he said, and the cheese is soft because the atoms that make them up are either closely or loosely packed. He boldly stated: "Nothing exists except atoms and empty space; everything else is opinion." He rightly believed that atoms were small, numerous, and virtually indestructible.

He went on to claim that the Universe was originally made up of atoms that randomly flew around until they bumped into each other, coalesced, and formed planets, like the Earth. Furthermore, he said that the Earth is a giant sphere flying through empty space and that our Universe is just one of many. It is truly remarkable when you consider that he was making these claims nearly 2,400 years ago. Unfortunately, however, Democritus and his followers had no proof of the existence of atoms or any real support for these other theories. They were philosophical speculations, no more.

Above:
Democritus' cosmology of the Universe put the Earth and the planets at the centre, surrounded by stars and an "infinite chaos" of atoms.

Democritus lived till the age of 90, by which time he was blind. He is then said to have starved himself to death. But he lived long enough to realize that he had lost out in the battle of ideas to those of the younger and more persuasive Aristotle, who, building on the ideas of Plato, claimed that all matter was made of five elements: earth, air, fire, water and aether. These were controlled by "forces" of love and strife. Unlike Democritus's atomic theory, which suggested that all life, colour, and variation in the world could be explained by the interaction of tiny particles, Aristotle's theory gave things purpose. This idea that the world was not random but imbued with purpose was for some a more comforting way of understanding life. Perhaps it was because of this that the so-called atomic theory was abandoned for over two thousand years. Aristotle became such an influential thinker that his views on this and many other areas of science (a great deal of which were wrong) not only prevailed in Greece, but were later taken up by the Islamic world and became the cornerstones of alchemical and medical practice in Europe until well into the Middle Ages.

ALCHEMY

The question "what is the world made of?" now became an obsession of the alchemists. The earliest form of alchemy was developed in China well over two thousand years ago. It was based on an understanding of the world that was similar to Aristotle's. Differing proportions of fire, wood, metal and water explained why matter had different properties. Unlike the Greeks, who enjoyed philosophical speculation for its own sake, Chinese alchemists focused their efforts in very practical directions. Encouraged by a succession of emperors who feared growing old and dying, the Chinese alchemists set out to find the elixir of life – described by some as drinkable gold. Ironically, one of their early discoveries was gunpowder, which rapidly became an elixir of death. By the 12th century the recipe for gunpowder, along with many alchemical beliefs, spread to Europe where they would be developed and go on to shape the history of the world.

GUNPOWDER

Gunpowder is a mixture of sulphur, charcoal, and potassium nitrate (saltpetre), mixed roughly in the ratio 2:3:15. It is widely believed to have been invented by Chinese alchemists or Taoist monks sometime around the 9th century AD, though there are competing claims. The Chinese rapidly realized its military potential and used it to create bombs and fuel rockets that they used against the invading Mongols, though it was clearly not a decisive weapon as the Mongols successfully conquered China and founded the Yuan Dynasty at the start of the 13th century. The Arabs probably learnt the secret of gunpowder from Chinese traders and in the latter half of the 13th century Hasan al-Rammah put together a book of gunpowder recipes in his *Book of Military Horsemanship and Ingenious War Devices*. About the same time, the Europeans learnt of gunpowder, possibly from the Arabs. Roger Bacon, an English friar and philosopher, wrote: "By only using a very small quantity of this material much light can be created accompanied by a horrible fracas. It is possible with it to destroy a town or an army."

THE 17TH CENTURY

Let us now jump forward to 1695 and the laboratory – if it can be called that – of Hennig Brand, one of the last of the alchemists. Brand believed he was finally on the brink of discovering the legendary Philosopher's Stone, a sort of supernatural universal cleanser. Alchemists like Brand believed fervently in the Philosopher's Stone. It was said that the Stone could clean out the impurities from base metals and turn them into gold, the purest metal of them all, as it never corroded. Similarly, the Stone could clean the impurities from a person, ridding him of disease, making him immortal and pure of soul. It is understandable then, that so many men throughout the ages, including men of undoubted genius such as the physicist Isaac Newton, dedicated so much of their lives to finding it. Brand was no such genius but he was extremely determined. Having already exhausted the fortune of his first wife in his quest for the Stone, he began working his way through the even larger fortune of his second wife.

Brand decided he might find the Philosopher's Stone in human urine. It seems an unusual place to look and sadly we do not know what persuaded him to try. Alchemists like Brand believed that they were recovering lost secrets – fragments of knowledge about how to change the world, dating back to the ancients. They were obsessed with codes and ciphers, not only as a way of protecting their professional position, but also for keeping potentially dangerous secrets to a limited number of individuals. So although we know some of what Brand did, much remains a mystery. Perhaps he chose urine because it was gold coloured and readily available, or perhaps because it had already been found to have useful and unusual properties. The Romans used it to wash clothes – their launderettes, known as fulleries, stank of urine, which they had discovered was excellent for removing grease. Urine has also been shown to be a useful way of producing saltpetre, an essential ingredient of gunpowder. A 13th-century recipe claimed that the best source, or that which would produce the biggest blast, was Bishop's urine.

> *"Alchemists like Brand believed they were recovering lost secrets – fragments of knowledge about how to change the world, dating back to the ancients."*

Above:
"Quicksilver", a metal that is liquid at room temperature, was associated with the Greek messenger god Mercury and was often represented by his symbol, the caduceus or snake-entwined rod.

Left above:
A page from a medieval work listing the supposed alchemical properties of various substances.

What we do know is that Brand collected several barrels of urine from the local soldiers and left it to ferment; he estimated that he would need 5,000 litres (1,100 gallons) for his experiment. Using a complex series of glass flasks, pipes, and ovens capable of achieving high temperatures, he reduced and distilled it, then added sand – perhaps reasoning that it also had a golden colour – and distilled it again. Those who have attempted to repeat Brand's "experiment" have come to appreciate what a very able technician he must have been. To be able to boil, purify, and extract anything using such crude heat sources and under such dangerous conditions was

Left: Entitled "The Alchemist in Search of the Philosopher's Stone", this spectacular painting by Joseph Wright of Derby (1734–1797) captures alchemy's strange blend of mysticism and scientific inquiry.

a truly remarkable achievement, let alone the smell. The white, waxy goo that remained after so much work must have been, initially, rather disappointing. There was no gold here, and certainly no Philosopher's Stone. But the goo did have a more remarkable property: it glowed in the dark. Brand named it "phosphorus", from the Greek meaning "light". Before long, Brand was touting phosphorus as an aphrodisiac and medical panacea, especially suitable for the treatment of mental conditions. Phosphorus appeared in 18th-century pharmacopoeias as a useful treatment, until it was recognized as being highly poisonous. It was then labelled "the devil's element".

We now know that Brand had isolated an element. He had made a significant discovery, but lacked any proper explanation of what was happening; he had the technology, and the methods, but not the science.

Initially, Brand had a great deal to be pleased about. Phosphorus was such a rarity that it was worth more than its weight in gold. An ounce of phosphorus sold for around six guineas. Not surprisingly, Brand tried to keep his production methods secret and for a while succeeded. But the volumes of urine required and the smell that the process produced meant it was inevitable that others would soon catch on.

"Phosphorus was such a rarity that it was worth more than its weight in gold."

MAKING PHOSPHORUS

It is ironic and tragic that phosphorus, which was first discovered in Hamburg, was used by the Allies in the bombs that later destroyed the city in World War II. The exact details of Hennig Brand's technique for producing it are not clear as he kept much of what he did secret, but broadly it consisted of leaving the urine to ferment and then boiling it down to a thick liquid or paste, which was then heated to a very high temperature, distilled, mixed, and heated again, until finally it yielded phosphorus. Brand needed around 1,000 litres (220 gallons) of urine to produce just 60g (2oz) of phosphorus. As it turns out, leaving urine to ferment was both odorous and pointless – just as much phosphorus can be obtained from fresh urine. His production method was also enormously inefficient, since one of the products he discarded after distillation contained much of the phosphorus present in urine. Brand's 1,000 litres (220 gallons) should have yielded over 20 times more phosphorus, since a single litre of human urine actually contains around 1.4g (0.5oz). Despite the cost and unpleasantness associated with its extraction, human urine remained the main source of phosphorus until the 18th century, when Carl Scheele, a Swedish chemist, showed that it could be extracted from bone ash.

Robert Boyle, a London doctor and the younger son of the Earl of Cork, was one of the first to uncover Brand's secret. He was told that phosphorus was derived from "somewhat that belonged to the body of man" and drew the correct conclusions. He made his phosphorus from urine, but unlike Brand realized its wider potential. Rather than poisoning his patients, Boyle used phosphorus to create matches. These were a great invention; the first reliable way to start a fire without need for flints to be cut, sticks to be rubbed, or eternal flames to be tended.

Boyle was, however, no mere match maker. Initially a believer in alchemy, he became deeply sceptical about its claims and was determined to make it more "scientific". In 1661 he published *The Skeptical Chmyist*, a book that finally reintroduced the idea of atoms, or "corpuscles" as he called them, to Western thought. Among many other things, Boyle experimented with compressing air by pouring mercury into a U-shaped tube of glass, sealed at one end. He discovered a relationship between the pressure he created and the volume of the trapped air. This "law", now known as Boyle's law, said that doubling the pressure would halve the volume of trapped air. Boyle concluded that the best explanation for this was that air consists of tiny corpuscles and that by adding pressure he was forcing them closer to each other. This suggestion was, however, not taken up and once again the idea of the atom faded into the background of mainstream thought.

We may laugh at Brand's belief that anyone could ever extract an elixir of life from human urine, but the fact is that alchemists like Brand paved the way for scientists like Boyle. Alchemists developed techniques, especially in reducing, separating, and distilling, that would prove essential for developing modern chemistry and a better understanding of the nature of matter.

Left: An illustration of one of Robert Boyle's most famous experiments, in which he used a vacuum pump to show that the sound of a ringing bell could only be heard if there was air to carry it.

THE CHEMICAL REVOLUTION

Moving on 80 years and by the middle of the 18th century gunpowder and cannons, discoveries of the Chinese alchemists, had been developed to truly terrifying levels of effectiveness by ingenious Europeans. Although the age of the great castle was over, brought to an end by the ability of cannons to shatter their defensive walls, religious differences fuelled almost continuous warfare across Europe and therefore the fate of nations still depended on the manufacture of cannons. The aim was to build bigger cannons, more cannons, and, crucially, cannons that would not explode in the operator's face. Artillery schools were set up in France and Britain to build a new class of technologically superior weapons.

"By the middle of the 18th century gunpowder and cannons had been developed to truly terrifying levels."

Making cannons required warring nations to develop their understanding of the metals needed to make them. This obsession with better metals sparked new interest in the processes of mining and turning ores into metal. Above all it brought about a new interest in gases, which was to be crucial in the next stage of attempts to understand matter.

HOW TO MAKE A CANNON

Although there is some evidence of types of cannon being used in the ancient world, the first that resemble what we would recognize today as a cannon almost certainly originated in China. They were simple tubes made of paper and bamboo, which were filled with gunpowder (another Chinese invention) and odd bits of shrapnel, and were almost as dangerous to the people who used them as they were to their enemies. Later, they were made of metal, iron, or brass. The Chinese mounted thousands of them on the Great Wall of China in a futile attempt to keep the Mongols at bay. From the Chinese, cannon technology seems to have been picked up by the Islamic world and then the Europeans. The arrival of the cannon in Europe transformed warfare, in particular the way that sieges were conducted. As Niccolò Machiavelli, a wily Italian political philosopher commented, "There is no wall, whatever its thickness, that artillery will not destroy in only a few days." By the time he was writing, in the early 1500s, people had realized that the longer the cannon barrel, the further it could fire. The result was that people started to build cannons that were truly enormous; cannons with barrels 3m (10ft) long and weighing over 9,000kg (9 tonnes) were not uncommon. As with the early bamboo cannon, however, there was considerable risk that the barrels would explode, killing the gunners.

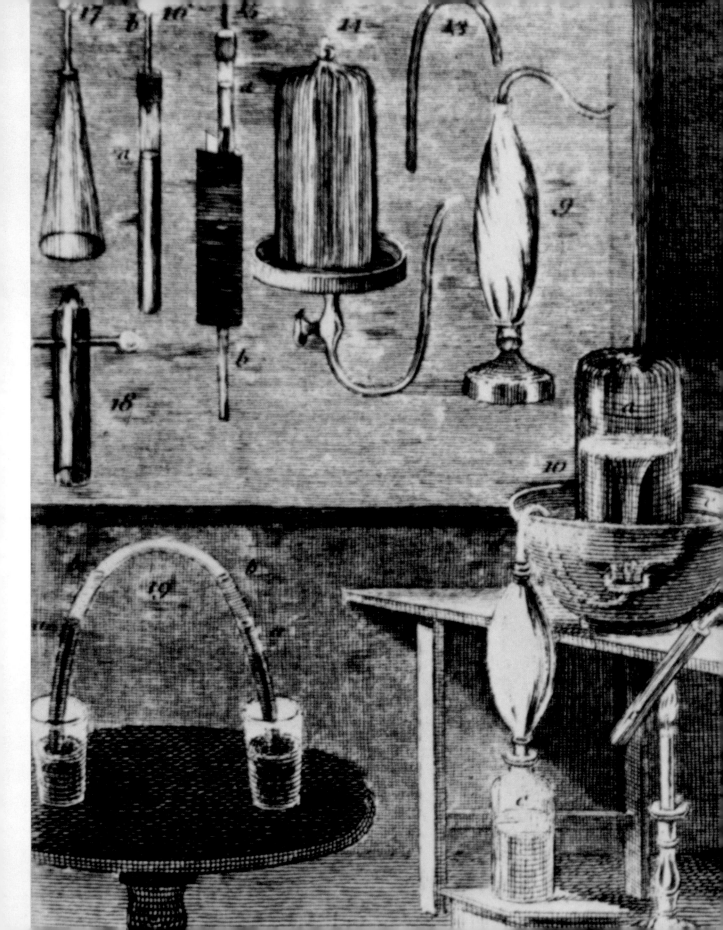

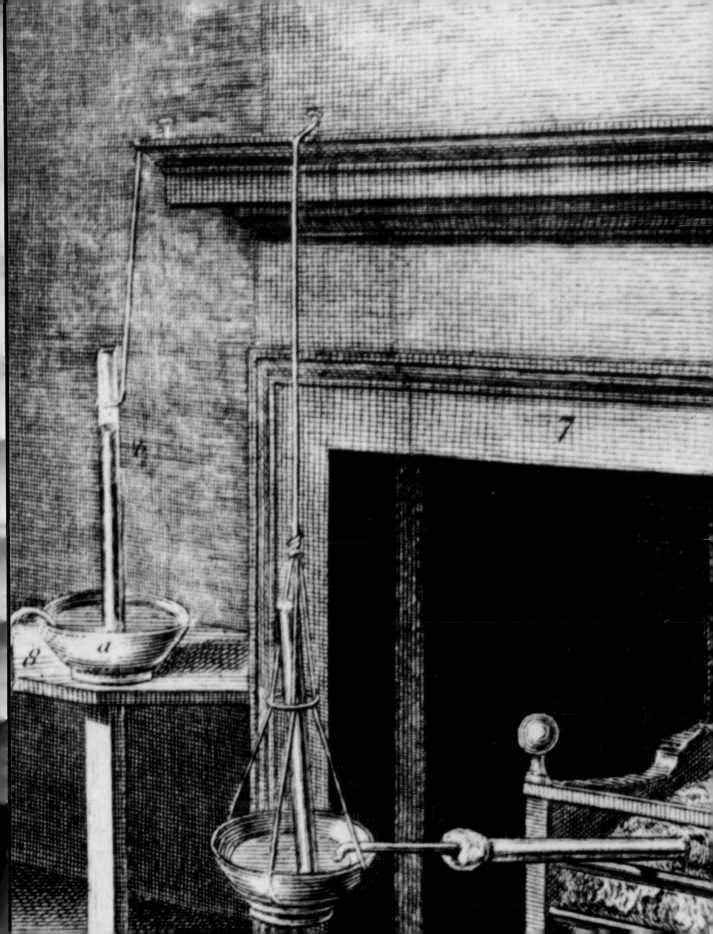

The nearest alternative natural colour to Perkin's mauve was purple. It was a colour that had been used to dye the cloaks of Roman Emperors and was enormously expensive, since the only way to make it was from the glandular mucus of thousands of molluscs. Perkin's dye was far cheaper to make and in many ways superior – it always produced the same uniform shade, didn't smell of fish, and, most importantly, didn't fade in daylight. Perkin's mauve became all the rage. In 1858, Queen Victoria wore it to the wedding of her daughter, the Empress Eugenie, who was a fashion icon. Soon, the streets of London were awash with people wearing mauve – an outburst of "mauve measles". It brought colour to the Victorian age, and a knighthood to William Perkin. Mauve dye opened the floodgates to a surge of new colours, and a new industry was created to produce them. Dyes were one of the first chemicals to be made on an industrial scale and others, including fertilizers, soap, and dynamite, quickly followed. The manmade, the synthetic, took over from the natural.

> *"Perkin's mauve became all the rage. Soon, the streets of London were awash with people wearing mauve."*

The man who had inspired Perkin was the German, von Hofmann, and soon his former colleagues seized the chemical initiative. Trained German chemists came over to England, learnt Perkin's secrets and returned home. By 1878, the products of coal-tar production in England were valued at £450,000, while those of Germany topped £2 million. German chemists, working within a university system that actively encouraged research, had discovered a whole new range of synthetic colours. From being importers of natural dyes, Germany became the world's leading exporter of synthetic colour.

Another important German innovation was Fritz Haber and Carl Bosch's invention of a way to "fix" nitrogen from the air. This enabled large-scale production of ammonia, which led to artificial fertilizers. The resulting boom in food production led to massive population growth. Today much of the world would starve if it were not for artificial fertilizers. But artificial nitrates could also be used for explosives, and soon were. Meanwhile, the dye manufacturing process was producing large amounts of toxic chlorine gas as a by-product, which was of interest to the military for a different kind of weapon – poison gas.

Left: William Perkin made his great discovery of "mauveine" dye (below) aged just 18. He continued to develop new dyes and even synthetic perfumes throughout his life.

Right: A bottle of Perkin's original mauve dye, preserved in London's Science Museum.

ORIGINAL MAUVEINE PREPARED BY DR WILLIAM PERKIN IN 1855

MYSTERIOUS RAYS

Henri Becquerel
1852–1908

Inspired by Röntgen's discovery, a French physicist called Henri Becquerel decided to investigate other phosphorescent materials to see if any of them also produced X-rays. His grandfather had been a famous collector of crystals that glowed in the dark, among them some uranium salts, so Henri used these for his experiments. He first "charged" the uranium crystals by exposing them to sunlight, then left them under some carefully wrapped photographic plates to see what would happen. Sure enough the plates subsequently showed the mark of the crystals. He repeated this experiment several times, then decided to put a copper cross between the crystals and the photographic plate. On previous occasions he had always charged the crystals with sunlight before doing the experiment. This time, either deliberately or because Paris was uncharacteristically gloomy, he did not. To his surprise the photographic plate showed the mark of the cross. The uranium salts were emitting mysterious rays, akin perhaps to X-rays, without needing to be charged.

Becquerel was clearly not that impressed with his findings, because he handed it over to one of his students, Marie Curie. Working with her husband Pierre, Curie soon discovered that pitchblende, the mineral ore from which uranium comes, produced huge amounts of energy without apparently losing mass. This was deeply puzzling because it seemed that these rocks were producing energy from nothing. In time it became clear that the pitchblende was unstable and was decaying to release different forms of radiation and huge amounts of energy. For her pioneering work Marie Curie won two Nobel prizes. She was also exposed to so much radiation during her research that it ultimately killed her. Her laboratory notebooks remain intensely radioactive, so much so that as historical artefacts today they have to be kept in lead-lined boxes.

Above: Working over four years, the Curies refined more than a tonne of pitchblende ore to extract just one tenth of a gram of a new element, radium.

Left: Marie and Pierre Curie, photographed at work in their laboratory at the School of Industrial Chemistry in Paris.

While Röntgen was investigating the mysterious X-rays produced by his Crookes tube, the physicist J.J. Thomson, working at the Cavendish laboratories in Cambridge (laboratories named after a relative of Henry Cavendish, the discoverer of hydrogen), had been studying Crooke's radiant matter. He agreed that the rays were made up of a stream of tiny particles, but his experiments suggested these particles were far smaller than the yet to be discovered atom. "The assumption of a state of matter more finely divided than an atom is a startling one," he declared. Thomson had found evidence of a subatomic particle – the electron.

THE ATOM

By the beginning of the 20th century it was becoming clear that the atom was not the smallest and most indestructible unit in the Universe, as people had been claiming. The Crookes tube and the discovery of radioactivity clearly pointed to the existence of particles that were far smaller than atoms. The problem was: how to study something so small that it could not even be seen. Working with two colleagues, Hans Geiger and Ernest Marsden, New Zealander Ernest Rutherford decided to use the atom to investigate the atom.

Rutherford had discovered that radioactive materials, as they decay, produce two very different types of radiation: alpha and beta particles. He used alpha particles, the heaviest and least penetrating, as his probes. Being notoriously clumsy, he delegated the actual experimental work to Geiger and Marsden, who spent days on end in a darkened room firing alpha particles at a thin sheet of gold foil. They could see where the alpha particles ended up by using a sheet of zinc sulphide, which sparkled when the alpha particles hit it. Most of the time the alpha particles passed straight through the gold sheet, but every so often a particle would be deflected. It was, said Rutherford, as surprising as if he had fired a 38cm (15in) shell at a sheet of paper and the shell had bounced straight back. Rutherford concluded from this that an atom consisted largely of empty space, with a small nucleus containing most of the mass. The positively charged nucleus, he said, was surrounded by tiny negatively charged particles – the electrons.

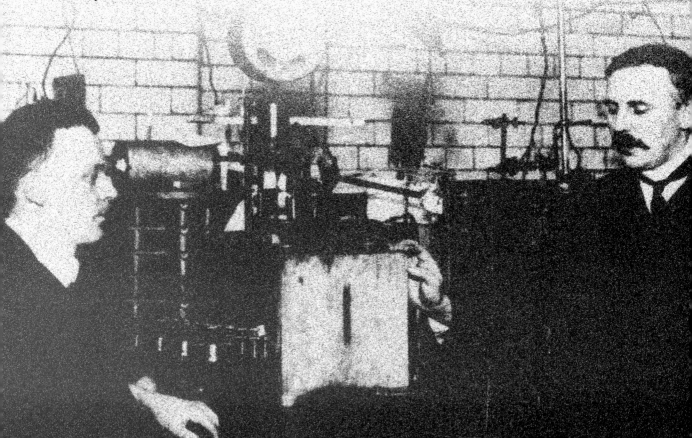

FLY IN A CATHEDRAL

To give some sense of scale, Rutherford imagined an atom blown up to the size of the Albert Hall. The nucleus, where almost all the mass of the atom resides, would be the size of a gnat – or as the newspapers put it, finding the nucleus was like trying to find a fly in a cathedral. This picture of a tiny fly buzzing around a vast cathedral is compelling but, in fact, quite wrong. Sticking with the example of a cathedral, the nucleus would actually be smaller than a grain of sand. Where Rutherford was right, however, was in claiming that the atom is mostly empty space. We, and everything in the world around us, consist almost entirely of void, of nothingness.

The mental picture that many of us have of an atom is that of a mini Solar System in which the electrons are like planets travelling through huge regions of empty space, revolving around the Sun, or the nucleus. This image makes sense and this was also how Rutherford first imagined the atom. But as other physicists soon pointed out, it was impossible that the atom could be anything like this. Unlike the planets, electrons carry a negative charge and classical physics predicts that a charged particle, when accelerating, will emit energy. If the electrons emitted energy, then they would soon collapse into the nucleus. But they clearly do not.

"We, and everything in the world around us, consist almost entirely of void, of nothingness."

A colleague of Rutherford's called Niels Bohr became so obsessed with the problem that he spent his honeymoon working on it. His solution was so radical in its implications that many of his colleagues found it impossible to accept. Bohr himself once claimed that anyone who is not outraged when they first hear about quantum theory has not understood it. What Bohr said was that electrons only travel in fixed orbits; they are more like trams than buses. Every so often an electron will jump into an orbit that is closer to the nucleus and as it does so it emits a discrete package or quanta of energy. This is a so-called quantum leap. In fact, the word "leap" is misleading because what is being described is more like teleportation.

Left: Hans Geiger (far left) and Ernest Rutherford with their experiment in the laboratory at the University of Manchester.

Right: Simplified illustration of the Bohr model of the atom, with electrons moving on discrete and well-defined orbits around the nucleus.

Others rapidly built on Bohr's model, creating a description of the atom that is so far from common sense that it is hard for most of us to grasp. Within the world of the atom physicists had encountered an area of the Universe that our brains are just not built to understand. In the weird world of quantum mechanics, as it came to be known, electrons can be described

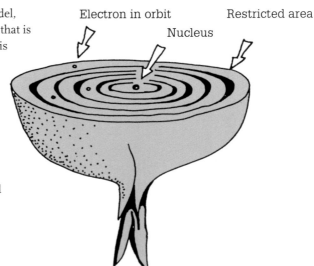

Electron in orbit   Restricted area

Nucleus

transistors on a chip would double every two years – Moore's Law. He was right. Intel, one of the world's leading transistor manufacturers, estimates that well over 10 quintillion (1 followed by 19 zeros) transistors are shipped every year. This equates to over a billion transistors for every human on earth.

William Shockley left Bell Labs in an atmosphere of bitterness and rancour after winning a Nobel Prize in 1956 and headed off to an orange grove near San Francisco. His reputation allowed him to recruit some of the finest minds in America, but soon they couldn't bear to work with him. Key men left the company but stayed on in San Francisco – and turned the orange grove into Silicon Valley, the high-tech heart of California.

SEMICONDUCTORS

Semiconductors lie at the heart of the transistor and therefore all modern electronic goods, such as computers, mobile phones, and televisions. The key to a semiconductor is the ability to control the rate of flow of electrons through it. Materials vary in how well they conduct an electrical current; metals, for example, tend to be good conductors, whereas rubber and plastics tend to be the opposite – good insulators. A semiconductor is a substance whose electrical resistivity – how strongly it resists an electrical current – lies somewhere between that of a conductor and that of an insulator. Critically, its resistivity can be modified by adding impurities (doping). Silicon, the most common element in the Earth's crust after oxygen, making up nearly a quarter, is widely used in the manufacture of semiconductors. This is why we talk about silicon chips and Silicon Valley, the place in San Francisco where transistor technology first took off.

Transistors are really just on–off switches. The semiconductor material in a transistor allows an electrical current to travel in one direction and control the size of that current. Running a small current through the transistor changes the electrical conductivity of the semiconductor, which then alters how much current passes through it. In this way, a small current can be used to control and switch much larger currents.

Above: A technician working on a detector at the European Organisation for Nuclear Reseach (CERN) laboratories, deep underground on the borders of France and Switzerland near Geneva.

THE FUTURE

Discovering the elements inspired a chemical revolution, which led to the development of medicines, plastics, and synthetic materials of every kind. Going down a layer took us to the atom. And probing the world of the atom has led, among other things, to nuclear energy, molecular biology, genetic engineering, molecular medicine, the computer, the laser, and the maser. It is estimated that the products of quantum theory account for about a third of the gross national product of industrialized countries.

Currently scientists at places like CERN, the European Organization for Nuclear Research and home of the Large Hadron Collider, are smashing particles together at extraordinary speeds in the hope of understanding the next level down – the quarks, which may or may not turn out to be the true building blocks of matter. A useful side product has been the World Wide Web, developed at CERN in 1989. The truth is we still don't know what the world is made of and it is possible we never will. The search, however, has been wonderfully productive.

Connections – Matter

Democritus
c460 – 370BC

Robert Boyle
1627 – 1691

John Dalton
1766 – 1844

REFORMATION

CLASSICAL GREECE

ROMAN EMPIRE

MIDDLE AGES

ISLAMIC SCIENCE

AGE OF DISCOVERY

RENAISSANCE

^ Crookes' tube

As is often the story, the Ancient Greeks hit on both the right and wrong answers – and it was the latter that held sway. Democritus' theory that everything was made of tiny particles called "atoms" was eclipsed by Aristotle's idea of the five elements: earth, air, fire, water, and aether. This belief was echoed in the medieval practice of alchemy – the semi-mystical search to transform matter.

Driven by research into gases, Robert Boyle and later John Dalton attempted to reintroduce the idea of atoms, without success. Meanwhile, however, a revolution was occuring in chemistry – first in understanding and isolating elements, then in manipulating them to synthesize new materials.

m Crookes
2 – 1919

JJ Thompson
1856 – 1940

Ernest Rutherford
1871 – 1937

Niels Bohr
1885 – 1962

EARLY 20TH CENTURY

AGE OF ENLIGHTENMENT

MID 20TH CENTURY

21ST CENTURY

ˆ First x-ray   ˆ Atom bomb   ˆ Atom model

By the end of the 19th century, most chemists agreed that atoms existed, but they thought that each element had its own unique atoms and was indivisible. The breakthrough began with investigations into the mysterious glow of William Crookes' electrified vacuum tube and the discovery of x-rays. J.J. Thompson realized that the rays were made up of particles even smaller than atoms: electrons. Rutherford built on this insight to define the structure of the atom, including the startling fact that it is mostly empty space. His colleague Niels Bohr solved the remaining problem of how electrons orbit the nucleus, opening up the head-spinning world of quantum mechanics.

Understanding the structure of the atom paved the way for splitting it – nuclear fission. Rutherford's first partial success with this happened in 1919; by 1932 he had succeeded and the countdown to the atomic age had begun.

Life

HOW DID WE GET HERE?

In the late Middle Ages it was believed that the lion was the King of Beasts – its Latin name, *leo*, meant literally that. It was also believed that it used its tail to rub out its footprints and to deceive hunters, that it slept with its eyes open and that its cubs were born dead, only to have life breathed into them by their father after three days – just as Christ was said to have lain dead for three days before rising to heaven.

All this knowledge could be found in the "Bestiary", a list of the world's animals – initially all 40 of them – first compiled by an anonymous Greek after the 1st century AD, and added to occasionally over the centuries as it was studied by Arab and Christian scholars. Copied and re-copied, with beautiful illustrations that became ever more unrealistic with every reproduction, by the 1600s it was the definitive source of knowledge about the animal kingdom. As the number of creatures included grew they were grouped into "families", but these also included mythical beasts that no one had ever seen, including the phoenix, the dragon, and the unicorn. All creatures were imbued with religious moral significance, and their existence declared as being according to the divine plan. There must be a sea horse because there were horses on land, and divine symmetry required there to be one in the ocean.

And where did we fit in? The question of how we got here was one that no one in Christian Europe really asked. We sat at the pinnacle of life, at the end of a great chain of being that stretched down through animals, insects, and plants to the lowest forms of life, with everything fixed in its place, and we had been put in that position by the Almighty. But then people began to have the courage to go out, to look, to draw, and to study the natural world as it really was, and it was this that led inexorably to the realization that the question of "how did we get here?" was one that needed to be answered.

Left: A 14th-century manuscript illustration blends medieval interpretations of real animals such as elephants with fantastical figures such as wild men and dragons.

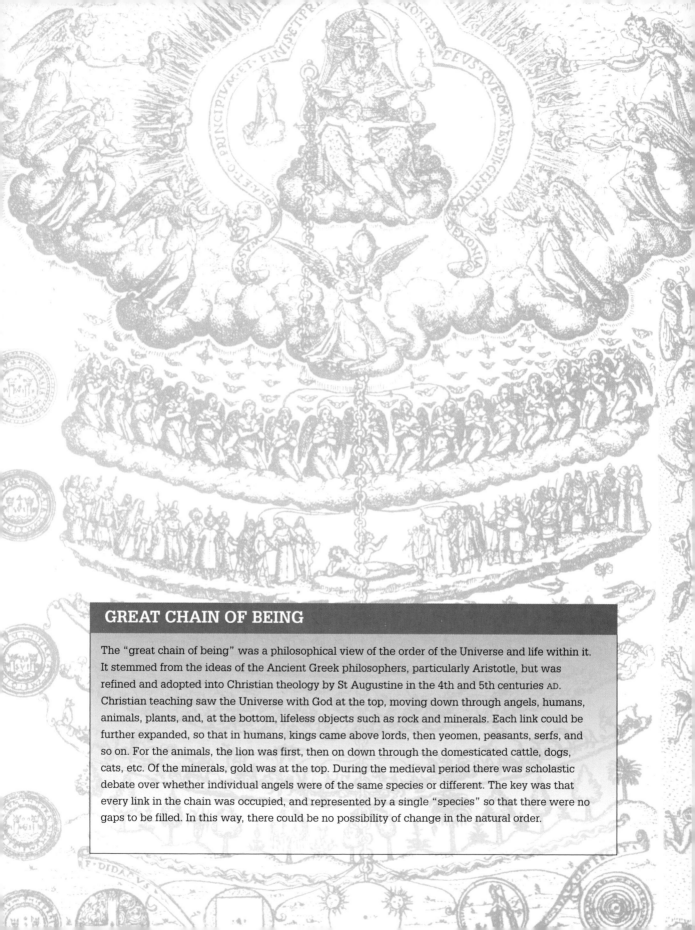

GREAT CHAIN OF BEING

The "great chain of being" was a philosophical view of the order of the Universe and life within it. It stemmed from the ideas of the Ancient Greek philosophers, particularly Aristotle, but was refined and adopted into Christian theology by St Augustine in the 4th and 5th centuries AD. Christian teaching saw the Universe with God at the top, moving down through angels, humans, animals, plants, and, at the bottom, lifeless objects such as rock and minerals. Each link could be further expanded, so that in humans, kings came above lords, then yeomen, peasants, serfs, and so on. For the animals, the lion was first, then on down through the domesticated cattle, dogs, cats, etc. Of the minerals, gold was at the top. During the medieval period there was scholastic debate over whether individual angels were of the same species or different. The key was that every link in the chain was occupied, and represented by a single "species" so that there were no gaps to be filled. In this way, there could be no possibility of change in the natural order.

CLASSIFICATION

Left: An illustration of the Great Chain of Being published in the *Rhetorica Christiana*, an account of Franciscan missionary Diego de Valades's evangelical journey through Mexico.

Sloane's passion for collecting did not stop when he left Jamaica; in fact this was just the beginning. On his return to England, he began to acquire objects of an astonishing variety, even whole collections from others. By the end of his life, income from chocolate, his upper-class clients, shrewd investments, and marriage into a wealthy family had made him a rich man and helped him to amass the largest collection of curiosities in the world. There were 265 volumes of pressed plants; 12,500 more "vegetables and vegetable substances"; 6,000 shells; 9,000 invertebrates; 1,500 fishes; 1,200 birds, eggs and nests; 3,000 vertebrates, both skeletons and stuffed samples; and human "curiosities". There were minerals, rocks, and fossils in their thousands, jewellery, ornaments, medals, coins, art works, ethnographic objects, and a library of 50,000 books. The collection was so large and so special that, six months after he died in 1753, the government created the British Museum to house it. His collections formed the basis of the present day British Museum, Natural History Museum, and British Library, which have become three of the foremost storehouses of cultural, literary, and scientific knowledge in the world. And, true to his physician's calling, Sloane also provided the grounds for the famous Chelsea Physic Garden in London.

HOT ROCKS

The first objective, scientific attempt to date the planet can probably be said to be the work of a geologist and naturalist from France with aristocratic pretensions, Georges-Louis Leclerc, Comte de Buffon. Born plain old Georges-Louis Leclerc, he inherited a vast fortune from his mother – including the village of Buffon, near Dijon – and thereafter, from the age of 25, called himself de Buffon. His geological experiment was based on speculation by Isaac Newton in his *Principia* (see Chapter One) about the rate of cooling of comets, which Newton had observed sometimes fell into the Sun. Buffon had the notion that such impacts might throw bits of hot Sun out into space and thus that the Earth started life as molten material spinning around its star, gradually cooling to the point at which it could sustain life. He reckoned that if he could work out how long it had taken for the Earth to reach this cooled state, he could estimate the date of its creation.

Buffon had his own forge make a series of iron balls to represent the Earth, ranging from about 1cm (0.4in) in diameter up to about 15cm (6in). He put them into the hearth, one after the other, until they were red hot. Then he took them out and timed how long it took before they were cool to the touch. In a description of his experiment it is recorded that he "had resort to four or five pretty women, with very soft skins ... and they held [the balls] in turns in their delicate hands, while describing to him the degrees of heating and cooling." By timing a range of different balls he was able to extrapolate his results to a globe the size of the Earth. When Buffon performed this experiment using several different-sized balls he concluded that it had taken a total of 42,964 years before the planet could sustain life, and an overall figure of about 75,000 years to reach its present day temperature.

Make no mistake, this was radical. Such a figure was an unimaginably long time by the standards of the day, and it was a result that drew ire from theologians at the University of Paris. But when Buffon published his figure, in 1749, he was careful to avoid any direct challenge to scriptures, and his *A Theory of the Earth* formed the first part of his monumental life's work, a description of the whole of the natural world, *L'Histoire Naturelle*, in 44 volumes. This series of books laid out a vision of the Earth that had a totally new perspective. Buffon argued that both the Earth and the life on it had a history and that he had identified seven great epochs of that history, which conveniently formed a metaphor for the seven days of creation. Above all, though, he suggested that species could change, with time and by migration from one part of the globe to another. Yet Buffon's experiments on time were just the beginning.

Right: Appointed head of the Jardin du Roi (royal gardens) in Paris in 1739, Buffon transformed them into a centre for research, paving the way for successors such as Cuvier.

Above: A pair of bird illustrations from an 1829 edition of Buffon's *Complete Works.*

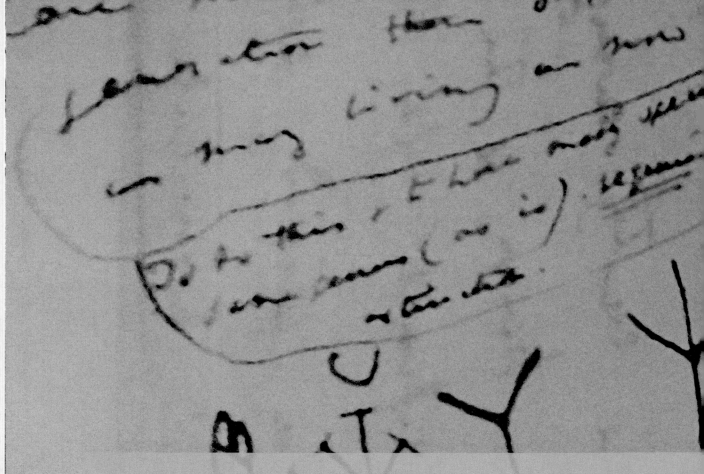

STRUGGLE FOR EXISTENCE

Charles Darwin
1809–1882

On his global journey Darwin experienced the same sense of wonder at the diversity of species that had impressed itself on Hans Sloane over a century before, and he witnessed firsthand the power of the Earth to create geological change when he experienced an earthquake in Chile. He was meticulous in his notes and careful as to the conclusions he drew from what he saw, but within two years of his return he had started to lay out his ideas on the "transformation of species". In one particular notebook from 1837 can be seen a sketch of a rudimentary tree of life, branching out from a single common ancestor – next to it, he wrote, "I think". As much as anything, his thinking was shaped by the time in which he lived. In Victorian Britain, industry and commerce thrived on competition; successful businesses made their owners rich, while those whose products were not good enough or cheap enough simply withered and failed. The apparently obvious success of Victorian capitalism was a clear indication that competition for survival was part of life. In those first years, Darwin was also influenced dramatically by reading, "for amusement", an essay by Thomas Malthus on the Principle of Population. Malthus argued that food supply could never keep pace with growing population, and that war, famine, disease, and poverty would always result, with the weakest falling into poverty and decline, and losing the struggle to survive. Politicians were using Malthus's logic to argue against support for the poorer working classes, for otherwise the British population could not be held in check. In essence, as Darwin himself later wrote, what he did was to extend Malthus's ideas of the

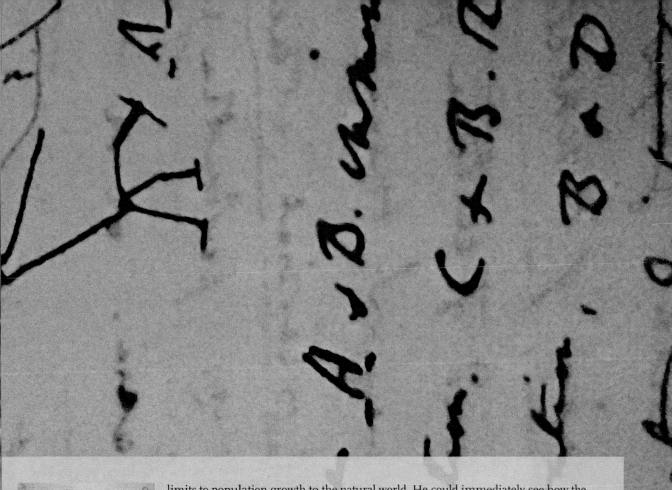

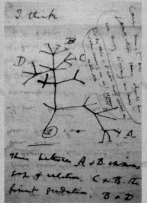

Above: Darwin's original sketch of a "tree of life". Unlike many of his supporters, Darwin did not make the mistake of creating a hierarchy among species.

limits to population growth to the natural world. He could immediately see how the possession of an advantage in the struggle for existence provided a mechanism by which species might continue to thrive, or face extinction.

With the elements of his theory in place, Darwin spent the best part of two decades amassing precise evidence to support it, and his caution was only underlined by seeing the furore that surrounded *Vestiges*. Here again, the nature of Victorian society shaped his progress. By the 1840s, natural history as a pastime had become hugely popular. As the railways began to carry more people to the leisure destinations of the seaside, so people took up the delights of studying nature. Many books were published on how to collect flora and fauna. Amateur naturalists abounded and as Darwin began to gather data he built a network of other gentlemen with whom he corresponded. Having settled with his new family at Down House, in Kent, and suffering recurring bad health, he remained a very private person, but wrote literally thousands of letters over his working life, exchanging information and samples – a practice made easier by the introduction of the Uniform Penny Post system. His network ranged far across the globe, with samples sent from countries throughout the British Empire, then still nearing the greatest extent of its global reach. Darwin carried out experiments on plants at home, studied barnacles for years and, crucially, became familiar with animal breeding. He kept and bred pigeons himself, and studied the effects that breeders have in selecting for specific characteristics in their pedigree animals, creating new varieties of dog, cat, cow, horse, plant, and pigeon.

Power

CAN WE HAVE UNLIMITED POWER?

In the last thousand years, human ingenuity has released vast quantities of energy from nature and created a world shaped around the on–off switch. The invention, design, and building of new machines and the discovery of successive new sources of energy are the backbone of this story. But it is a story of two threads that are intertwined: one tells of how we learned what power does; the other of how we discovered what power really is. Yet the search for new energy sources was led not by theorists trying to un-cover grand scientific laws, but nearly always by practical people on the make: the inventors, industrialists, and investors who saw how to exploit energy, and who dreamed the dream of unlimited power, and fortune. Attempts to reach that goal have not only helped to create the modern world but, when theory finally caught up with practice, also given us some of the most profound insights in the history of science – about the nature of time, and the fate of our Universe.

Left: Electrical power has enabled human beings to finally conquer the night, and created our modern 24-hour civilization – but the path to it has also revealed fundamental principles of physics and laws of nature.

INFORMATION FLOW

The world of electricity and the world of power were soon to become linked – although not in quite the way one might have expected. Oersted's finding caused a scramble among people with ideas to exploit the phenomenon, and what quickly emerged were several schemes to use the effect to create signals at a distance. By switching a current on and off it was possible to cause a needle to flick at the other end of the wire. Within little over a decade, several proposals for electromagnetic telegraph systems appeared in Germany, Britain, and America. Pavel Schilling, a Russian, demonstrated a telegraph of eight wires, with magnetic needles suspended on silk threads at the end, running between the rooms of his house; David Alter in Pennsylvania linked his house and his barn; and in Germany Carl Friedrich Gauss ran 1,000m (1,100yd) of wire over the rooftops of Göttingen. In Britain, a commercial telegraph was installed in 1839 along a section of the Great Western Railway, and six years later it enabled the police in London to be waiting for a runaway murderer who had boarded a train in Slough, some 40km (25 miles) away, when he arrived at Paddington Station. Here was a technology with an obvious practical benefit.

Perhaps 200km (125 miles) of railway lines could be travelled in Britain in 1830. By 1860, there were over 10,000km (6,200 miles). With the long straight tracks to run alongside, the telegraph wires rapidly transformed communication everywhere, but nowhere more so than in America. There, Samuel Morse's patented code, developed with his assistant Alfred Vail, became universally adopted, and as the great railroads carved their way across the continent, so the telegraph went with them. Thereafter, electricity has always been at the heart of information flow. The telephone, which converts sound to electric current and back, swiftly followed. Then came radio, which depends on the transmission of electromagnetic waves. And today's digital communication depends on the changing state of electrons.

Samuel Morse
1791–1872

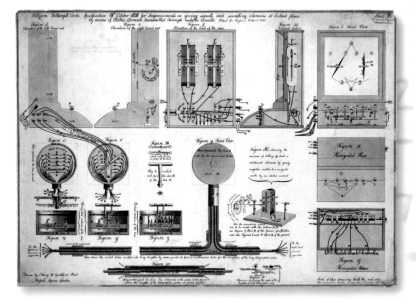

"The steam engine and the telegraph changed the world. Together they brought us the concept of universal time."

Left: Illustration from an 1838 patent for William Fothergill Cooke and Charles Wheatstone's first electric telegraph system.

The steam engine and the telegraph changed the world. Together they brought us the concept of universal time. Before it was possible to travel at the speed of steam there was no need for the time to be exactly the same in, say, London in the East of England and Cornwall in the South West. The world operated on a matter of hours, and it did not matter that Penzance, at the tip of the Cornish peninsula, ran eight minutes behind London. But now, precise time did matter. An eight-minute difference in the watches of two train drivers could easily result in a crash. The telegraph was quickly adopted by the railways; it was the only way to get a message to a station, or a signalman, before the train itself arrived. To ensure the whole network would run smoothly, all the trains began to adopt railway time – eventually referred to as Greenwich Mean Time. By 1860, almost every public clock in Britain showed the same time, and over the next few decades other countries around the world followed suit. In the USA, thousands of local "noons" gave way first to railroad times set by the headquarters of the rail companies and then to the standard time zones we know today.

With the telegraph, runaway criminals could be caught and runaway brides could be stopped before they took their vows; information moved at astonishing speeds not only across nations but across empires. By the 1870s, every continent was linked by cable, and keys tapped code from one side of the world to the other. The nature of the empires themselves was changed. Garrisons were needed to guard and secure the lines of cables and plantations of gutta-percha trees were grown to maintain the supply of their sap that formed the latex used in insulating the thousands of miles of electrical wire now threading around the globe. The fortunes of countries like Malaya, where the trees were grown, were transformed in consequence, as their value to the British economy changed. And above all, ever greater quantities of coal were needed to fuel the thousands of steam engines that now powered industry and transport.

THE TELEPHONE

The success of the telegraph meant that communication of sound at a distance became a clear goal. The invention of the telephone is usually credited to Alexander Graham Bell, a Scot living in America, who obtained the first patent for the device in 1876 and went on to commercial success. However, it is argued that the Italian Antonio Meucci demonstrated an early telephone some years before. Meucci had developed an electric "treatment" for rheumatism, and while electrocuting one of his patients he heard a scream seemingly passing down the copper wire – the sound vibrated an electrical conductor near the patient and created an electrostatic charge that in turn vibrated the electrical conductor near Meucci's ear. Meucci went on to develop a device with electromagnets linked to diaphragms. The principle was that sound moved the diaphragm, which moved the electromagnet, which in turn fluctuated the current down a wire. At the other end, the process in reverse recreated the sound. No working models of Meucci's device survived, and he failed to renew the payment for his patent application. It remains disputed as to whether or not he was the first inventor of the telephone.

EXPERIMENT AND PRACTICE

It was not long after Oersted's demonstration that electricity could move a magnet that the opposite relationship was seized upon. In London, Michael Faraday had been an apprenticed bookbinder, but had tried every possible approach to get a job working in some capacity in science, about which he was passionate. Chance events at the Royal Institution, including Humphry Davy temporarily blinding himself in a chemistry experiment, and the laboratory assistant being fired for assaulting the instrument maker, resulted in Faraday becoming Davy's new laboratory assistant. Faraday rapidly became Davy's key experimenter in electromagnetism and in 1821 managed to build a device that enabled an electric current to produce the circular motion of a magnet. This was the start of work that eventually led him to produce an electric motor. More significantly, ten years later, Faraday designed a machine that did the reverse: a moving magnet created an electric current in a coil of wire that was wound around it. This was the principle behind the dynamo – a constantly turning motor, such as a steam engine, could generate an electric current. What Faraday had done was to create both halves of the electrical power industry: electric current can generate movement; and movement can generate electricity.

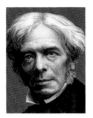

Michael Faraday
1791–1867

Dynamos and electric motors at first remained pretty much within the confines of the scientific laboratory or were used as curiosities to impress the public. But by the 1870s, a dynamo had been built that could supply electricity on an industrial scale, this time by a Belgian engineer, Zénobe Théophile Gramme, who also built a near identical machine operating on the reverse principle to act as a motor. With industrial-scale electricity now available, factories could use steam or water power to turn huge dynamos and generate electrical power, which they could then distribute to smaller electric motors to drive their machines, rather than the complex belts and drives that were needed to connect directly to steam engines themselves. By the end of that decade, the incandescent light bulb had also been successfully produced to a commercially viable standard by Thomas Edison, so that the workplace could now easily be lit well into the night. The new electrical technology ushered in what has become known as the "Second Industrial Revolution". Countries such as Germany, who had come late to industrialization, were now able to develop their economies rapidly, and the industrial and commercial power balance in Europe began to shift away from the dominance of Britain.

Right: An electrified factory at Hanover, Germany, in the early 20th century. The introduction of electric power was an important element of the so-called "Second Industrial Revolution".

Right: Electric motors use the repulsion and attraction between fixed magnets and the changing electromagnetic field in a coil of wire to turn a rotor and spin a drive shaft.

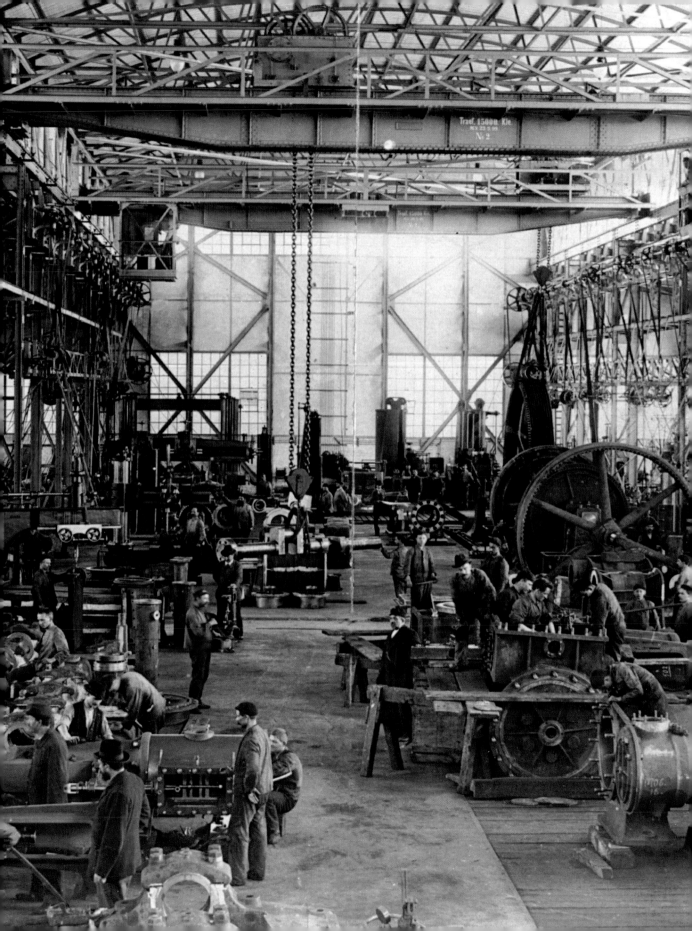

PERSONAL POWER

Yet by the end of the 19th century, it was still unthinkable that electricity would come to dominate the planet as it does today. For all the changes at the factory, walking home at night still meant walking through a world illuminated via the burning of coal gas. Electricity offered huge potential to change all that, but it suffered from one major limitation: distribution. It was all very well to have a steam engine next to a factory to turn the generator, to turn the motors, and light the light bulbs. But the energy lost along the wires was so great that it would require a steam engine and generator at the end of every street to service a town. Indeed, early distribution networks had a limited run of about 2km (1 mile). The challenge was to come up with a system that could offer useful electricity at the end of a very long wire.

In 1883, the Free Niagara Movement was triumphant in its campaign – one of the first environmental movements – to return Niagara Falls to a more natural state than was then experienced by every visitor to the area. The vast natural power source of the waterfall that lies on the Canada–USA border had been tapped since the first settlers cut mill races into the banks of the river a century or more before. But by now the shores on both sides were cluttered with mills and factories of every sort, all driven by the relentless flow of the water. The creation of a state reserve around the falls meant that the commercial free-for-all was brought to an end, but the bulk of the power of the falls remained untapped – clearly a terrible waste of commercial opportunity. Then, in 1886, an engineer on the nearby Erie Canal, one Thomas Evershed, proposed a huge engineering project to build a series of tunnels and channels to carry the power of the waterfall away from the reserve, so that it could be used. The potential cost was astronomical, and yet it was not obvious how the power could be exploited in order to get a return on such a huge investment. The Falls generate 8,000,000 horsepower, far too much for the needs of the nearby small town of Niagara Falls itself, which had a population of just 5,000, so the question was asked: could it somehow be distributed to the growing town of Buffalo (population 250,000 and rising) some 40km (25 miles) away?

Right: Up to 5,700 cubic metres of water pass through the kilometre-wide falls at Niagara every second, making them an irresistible source of power.

POWER STRUGGLE

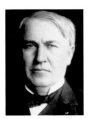

Thomas Edison
1847–1931

Two giants of the electricity industry provided two different answers. Thomas Edison, inventor of the first practical light bulb and many other electrical devices, had set up small local electrical distribution networks in Manhattan, using direct current. This is a form of electricity where the current flows continually in one direction, and it is subject to great loss due to the resistance of the wire along which it travels. To travel more than about 2km (1 mile) either the current at the start has to be so high as to burn out any light bulbs or motors, or the cables have to be too thick to be practicable. On the other side of the competition, George Westinghouse, inventor of the compressed air braking system for railways, had bought up patents belonging to Nikola Tesla, a disaffected former employee of Edison, who had come up with key improvements to the concept of alternating current – where the electric current literally flows in alternate directions along the wire. The idea behind this is that higher voltage is used to transfer the current for great distances, while a transformer – for which Tesla's inventions were critical – steps the voltage back down to a more manageable level at the consumer end.

The battle between these two cut-throat entrepreneurs was the high point of the "War of Currents" that had been running for over a decade, with different camps across Europe and the USA claiming their system was best. Edison tried every marketing trick in the book to discredit alternating current, including emphasizing its high voltage dangers, by deliberately promoting the use of Westinghouse's system for the newly invented electric chair. In the end, the first successful long-distance distribution of electricity was demonstrated in 1891 at the International Electro-Technical Exhibition in Frankfurt in Germany. The electricity was generated as alternating current at a cement works in Lauffen am Neckar, 175km (109 miles) to the south, and carried on overhead wires to light a display of a thousand bulbs at the entrance to the exhibition and, ironically, to power the turning of an artificial waterwheel. The spectacular success of this event was a deciding factor in the plan for Niagara, and it was alternating current that won the contract. At midnight on 16 November 1896, Tesla and Westinghouse's system proved itself; power from their transformers reached Buffalo. The first 1,000 horsepower went to the street railway company, and the local power company had immediate orders from residents for 5,000 more. Within a few years the number of generators at Niagara Falls grew to ten, and power lines were electrifying New York City. Broadway was ablaze with lights; the elevated street railways and subway system rumbled. Electrical power, the crucial element for nearly all modern technology, would soon be everywhere.

"Electrical power, the crucial element for nearly all modern technology, would soon be everywhere."

Left: Reliable electric lighting transformed cities in Europe and America, making the streets safer and helping to foster a 24-hour society.

Right: Edison's incandescent light bulb was not the first of its kind, but a number of innovations made it more robust and able to function within a power network.

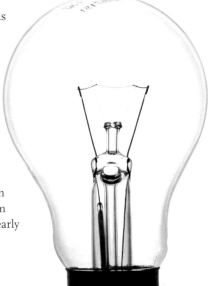

THEORY CATCHES UP

Theory did in the end catch up with the practical exploitation of power. Michael Faraday, as he studied the effects of electricity and magnetism, came to believe that there was a hierarchy of forces, with electricity – God's force – at the top, and gravity somewhere below. Faraday was the supreme experimenter of his age, and showed conclusively that electric current, electrostatic charge, and magnetism were at heart the same phenomenon. He argued that electricity and magnetism acted along "lines of force", taking time to move, perhaps in the form of wave motion, but he did not have sufficiently good mathematics to take his ideas of magnetic and electrical "fields" any further. Instead, it was a Scottish mathematician, nicknamed "Daftie" at school but regarded by many as the greatest classical physicist since Newton, who bound the ideas together as mathematically precise laws. James Clerk Maxwell was brought up in Edinburgh, attended university there and at Cambridge, and became a professor at Aberdeen at the astonishingly young age of 25, before settling at King's College in London. He worked on the physics of colour and the nature of the rings of Saturn, but his crucial work was to take Faraday's ideas of force fields and show mathematically that electricity, magnetism, and also light itself were manifestations of exactly the same phenomenon – electromagnetic waves. In 1888, nine years after Maxwell's death, this was proved conclusively when the German physicist Heinrich Hertz demonstrated the existence of radio waves, and showed that they travelled at the speed of light.

James Clerk Maxwell
1831–1879

Through the notion that the speed of light is always constant, Maxwell's equations also provided the basis for the radical ideas that emerged from the mind of Albert Einstein as his theory of special relativity. Published in 1905, it turned physics on its head, running counter to the existing Newtonian mechanical vision of the Universe and introducing counterintuitive concepts such as space-time, length contraction, and time dilation into the physicist's vocabulary. But it also provided a fundamental explanation of what power actually is, in the most famous equation in science: $E=mc^2$. Energy equals mass, multiplied by the square of the speed of light, an almost unimaginably large number. Until this point energy had been understood as heat, electricity, even

Below: The nuclear fusion reactions at the heart of a hydrogen bomb explosion unleash incredible destructive force – and properly harnessed, they may also offer a solution to our future energy needs.

wind or water power, and the different forms could be transformed into one another. But this was a far stranger level of equality; heat, motion, and radiation could all now be seen as expressions of this fundamental insight. Mass – the material of anything – contains almost incomprehensibly vast amounts of energy, which can be released by motion, burning, compressing, or splitting.

None of the great inventors or engineers of the previous three centuries had needed that equation to build the machines that built empires or transformed nations, but now that it was there it explained all their work. It explained too the awesome power that could be released by the splitting of the atom. Radioactivity suddenly could be seen as the raw energy released by infinitesimally small amounts of mass simply decaying. So for the first time, scientific theory set a goal for the inventors and engineers to aim for. Einstein himself, at first, was certain that such power could never practically be achieved, but of course we know now that in this he was wrong. The atomic bomb was created, and today's nuclear arsenals are testimony to the achievements of the science.

> **"Whatever form of energy we tap, it almost always comes to us via the Victorian technologies for generating and distributing electricity."**

There is more irony in the story of power. Today, nuclear radiation is a major source of energy for generating electricity, but the radioactivity in nuclear power stations is simply another fuel like coal, gas, or oil. It creates heat, to make steam, to drive turbines, which turn generators. This is true for most of the latest means of providing energy, be it renewable sources like wind or waves, hydroelectric, or even green fuels such as biomass. Whatever form of energy we tap, it almost always comes to us via the Victorian technologies for generating and distributing electricity.

So it was the search for limitless power that gave us the practical machines that drive the modern world. At the same time, the attempts by experimenters and theoreticians to find out how those machines worked revealed deep truths about science. They arrived at equations that reveal the vast potential of energy that exists in the world, and the laws of thermodynamics that determine why none of it can in fact be limitless.

CLASSICAL GREECE

ROMAN EMPIRE

ISLAMIC SCIENCE

MIDDLE AGES

AGE OF DISCOVERY

RENAISSANCE

REFORMATION

^ First battery

^ Leyden Jar

The story of creating power begins with the exploitation of the most obvious natural resources around us – water and wind. In the Age of Enlightenment, two other sources emerged. Electricity was a fleeting charge created by friction until in the mid 1740s Pieter van Musschenbroek invented a way of storing it, a capacitor or "Leyden jar". Alessandro Volta improved this still further with his "voltaic cell", the first practical battery.

At the same time the age of steam began. Primitive steam engines had existed for centuries, but Thomas

Alessandro Volta
1745 – 1827

James Watt
1736 – 1819

James Clarke
Maxwell
1831 – 1879

Michael Faraday
1791 – 1867

Thomas Edison
1847 – 1931

EARLY 20TH CENTURY

MID 20TH CENTURY

AGE OF ENLIGHTENMENT

21ST CENTURY

^ Early electric motor   ^ Lightbulb   ^ Telephone   ^ Atomic explosion

Two related inventions of Michael Faraday made electricity into a power source that would harness and ultimately eclipse steam – the electric motor and, more significantly, the electric generator. By the late 19th century, the "Second Industrial Revolution" began, with innovations such as the lightbulb (first reliably built by Thomas Edison) and the telephone changing the way we lived, worked and communicated for ever.

The last act of the story is where the overwhelmingly practical nature of the search for power connects with the theory, James Clerk Maxwell's research into electromagnetic waves paving the way for the work of Albert Einstein and the awesome power of the atom.

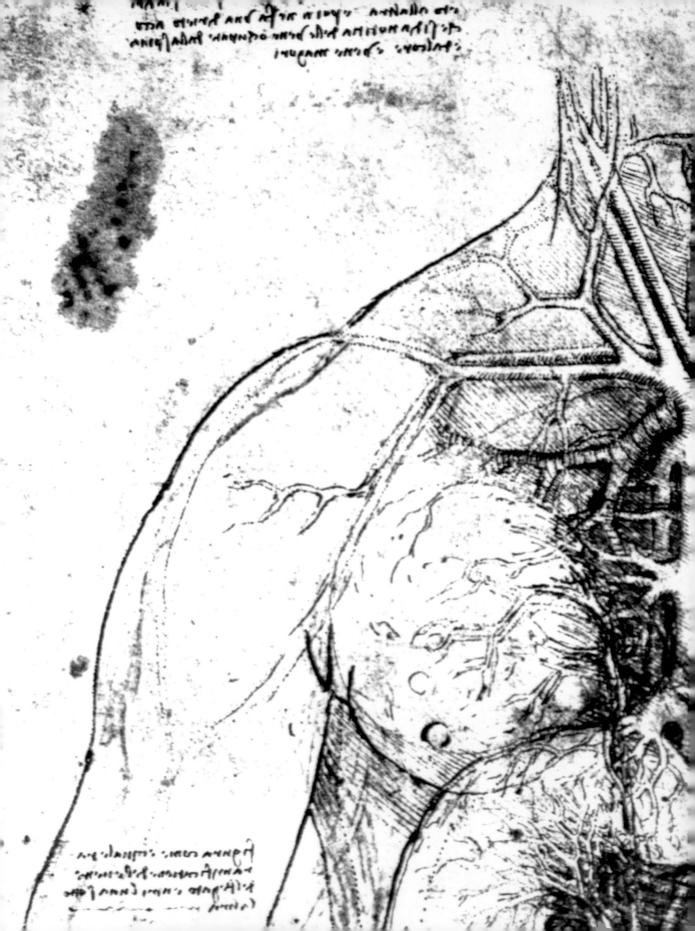

Body

WHAT IS THE SECRET OF LIFE?

Go out shopping and buy 18kg (40lb) of carbon, enough phosphorus to make two thousand matches, and a small iron nail. Then drop in on a friendly chemist and collect small amounts of a few other relatively common elements. Take it home, mix it in a bucket, add about 50 litres (11 gallons) of water, and stir. The resulting mixture is chemically similar to a person. Yet it is not a human and cannot, of course, be brought to life.

So what is the secret to life? What is it that turns a pile of chemicals into a living, breathing biological entity? The hunt for answers to these questions has created modern medicine and allowed us to reach a point where it seems that life itself will shortly be synthesized in the lab. The moment we create an entirely manmade cell is not that far away and when it happens it will be one of the most extraordinary in our history.

Over the last couple of centuries in the West, two rather different approaches have been employed in trying to find out exactly what it is that makes us tick. The first has been simply to cut things open and take a look inside. Is life, perhaps, simply a product of the way we are put together? This approach has been productive, but we may now have reached the limits of what further dissection can tell us. The second approach has been to look for a life force – something physiological that could explain the difference between a dead body and a living one. This approach, as we will see, led to important discoveries like biological electricity and the role of hormones, but in itself it did not bring us any closer to answering the question "What is life?"

Left: A detail from one of Leonardo Da Vinci's anatomical sketches of a human torso. Despite such detailed anatomical studies, figuring out the function of the various organs would be a long struggle.

THE ANATOMISTS

The first person to create accurate drawings of the human body was the Renaissance genius, Leonardo da Vinci. The illegitimate son of a lawyer and a peasant woman, Da Vinci was not given a formal education; perhaps because of this he was more questioning, more willing to rely on his own eyes than on the wisdom of the ancients. When he drew, he drew what he saw, not what others told him he should see. His paintings relied on new studies in perspective, but they also relied on a complete understanding of the human body. When he was first apprenticed, his master, the painter Andrea del Verrocchio, insisted that he and all his fellow students study human anatomy. Later, when he was a famous artist, he gained more intimate insights into the human body by dissecting corpses. At the time – the early 16th century – this was not illegal, but it was frowned upon and undoubtedly grisly. He dissected mainly at night, with the help of a young assistant, and was often in a race against time and the rapid onset of decay.

Da Vinci's anatomical drawings are amongst the finest ever produced and include the first drawing of a foetus in utero. But what is particularly striking about Da Vinci's work is that he not only drew what he saw but also reached insightful conclusions from what he observed. By comparing the arteries of an old man with those of a young boy, for example, he concluded that the furring up of the old man's arteries had contributed to his death. He had effectively described atherosclerosis hundreds of years before anyone else and at a time when many doctors thought that arteries carried nothing but air. Unfortunately, as with much of his work, he never finished it to his own satisfaction – there was always something else to find out about – and it was not published for another 160 years.

Leonardo Da Vinci
1452–1519

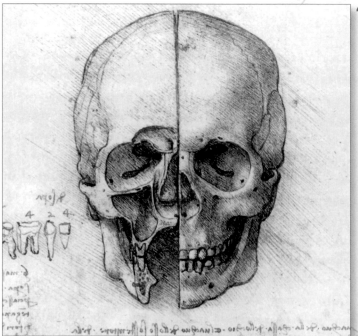

"Da Vinci's anatomical drawings are amongst the finest ever produced and include the first drawing of a foetus in utero."

Left: Around 1489, Leonardo completed a series of detailed drawings of the skull, attempting to locate the seat of specific mental faculties within the brain.

Instead, in Da Vinci's day, doctors relied for their understanding of human anatomy on crude drawings from the 2nd century by a man called Claudius Galenus, also known as Galen. A Greek, born sometime around AD 130 in what is now Turkey, Galen had begun his career treating gladiators, which must have given him an intimate knowledge of the human body in extremis. Because dissecting human bodies was actively discouraged at that time, his studies were confined to animals, from monkeys to pigs, and he took great pleasure in cutting open baboons in front of interested Roman citizens to show off his great anatomical knowledge. His textbooks were seen as almost sacred texts and never questioned. It was not till long after Leonardo's death that a Flemish dwarf called Andreas Vesalius would finally change that.

Right: Greek physician Galen was the ultimate authority on all medical matters for almost 1,500 years.

CHINESE MEDICINE

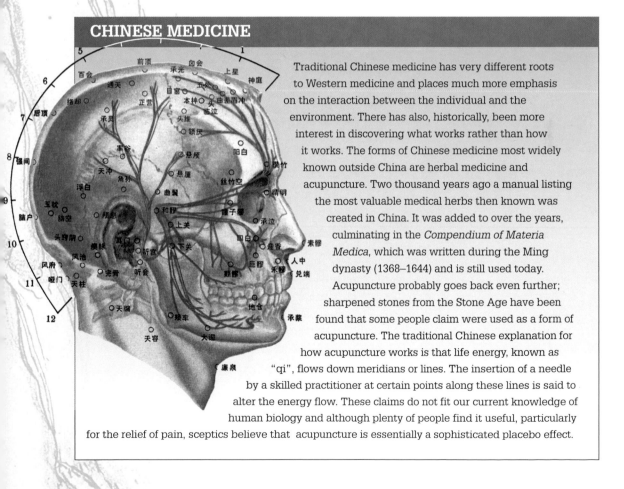

Traditional Chinese medicine has very different roots to Western medicine and places much more emphasis on the interaction between the individual and the environment. There has also, historically, been more interest in discovering what works rather than how it works. The forms of Chinese medicine most widely known outside China are herbal medicine and acupuncture. Two thousand years ago a manual listing the most valuable medical herbs then known was created in China. It was added to over the years, culminating in the *Compendium of Materia Medica*, which was written during the Ming dynasty (1368–1644) and is still used today. Acupuncture probably goes back even further; sharpened stones from the Stone Age have been found that some people claim were used as a form of acupuncture. The traditional Chinese explanation for how acupuncture works is that life energy, known as "qi", flows down meridians or lines. The insertion of a needle by a skilled practitioner at certain points along these lines is said to alter the energy flow. These claims do not fit our current knowledge of human biology and although plenty of people find it useful, particularly for the relief of pain, sceptics believe that acupuncture is essentially a sophisticated placebo effect.

DEATHLY DISSECTIONS

Picture the scene: it is 1536 and a criminal has been hung and left to rot on the gibbet. One evening, along comes Vesalius, a 22-year-old medical student. He looks up at the corpse with longing; human dissections may not be illegal, but body snatching most definitely is. That does not deter him. Vesalius jumps up, grabs the legs and pulls. With a terrible ripping sound they come off in his hands. He runs away into the night, clutching them in his arms. Later, he returns for the rest of the body.

What Vesalius was doing was extraordinarily dangerous. Not only was he risking jail and personal ruin, but his meticulous dissections of this and other bodies would challenge a belief system that had remained in place for centuries. At Vesalius's medical school there was just one set of anatomical textbooks – those written by Galen, whose words would be read out by the senior doctor while unquestioning students looked on, nodding, from a distance. Vesalius, however, had decided to do something that would have outraged and disgusted his contemporaries: to dissect and examine a human body himself.

Undeterred by the rotting human flesh, Vesalius put the stolen body on his kitchen table and set about stripping it down to its bare bones. He treated the task as if he was making beef stock. First, he filled a big pan with water and set it to boil. Then, he took bits of the corpse and removed as much skin and flesh as he could before dropping them into the pan, where they boiled away for hours, until falling apart. Finally, bone by bone, he tried painstakingly to identify every single part of the human skeleton. It was an arduous task; there are 206 bones in the human body, and this was just the start. Vesalius wanted to map not just the bones, but also every organ, ligament, and muscle, and was determined to understand where they all fitted – it was like putting a jigsaw back together. To do this he needed more bodies. So, naturally, he stole them.

Left: Vesalius's detailed anatomical drawings overturned many preconceptions that had remained unchallenged for over a millenium.

Below: Andreas Vesalius 1514–1564. Born into a family of physicians, the great anatomist spent much of his childhood dissecting birds, mice, and other small animals.

"Vesalius correctly identified the location of all the major organs, nerves and muscles in the human body and, significantly, began the proper study of human anatomy."

Above:
Illustrations from Vesalius's work, including the frontispiece illustration of a theatrical dissection (left).

Later, as chair of Surgery and Anatomy at the University of Padua, he found a sympathetic judge who was able to provide him with a supply of bodies obtained in more legitimate ways. Despite being a professor, Vesalius continued to carry out dissections himself and employed talented artists to illustrate his work. These drawings were eventually put together in 1543 in the form of a book, *De Humani Corporis Fabrica* (On the Fabric of the Human Body). In this book, he pointed out that much of what Galen had written was incorrect, and that 1,300 years of medical teaching were seriously flawed. Some dismissed him as a madman. Others, who took the time to read more carefully, wondered if perhaps the human body had changed since Galen's time. In all, Andreas Vesalius corrected over two hundred of Galen's mistakes.

Vesalius's masterpiece was published in the same year as Copernicus's great work, *De Revolutionibus* (see Chapter One). Thus, 1543 is regarded by some as the start of the scientific age, although this could be considered a little farfetched, since Copernicus's book would not have any significant impact on the world till long after his death. *De Fabrica* was not covering a subject that was quite as epic in its scale as the realignment of the Solar System, but it was in its own way a truly monumental piece of work. In it, Vesalius correctly identified the location of all the major organs, nerves, and muscles in the human body and, significantly, began the proper study of human anatomy. He had shown what he and others like Da Vinci passionately believed – that the wisdom of the ancients could not be relied upon and that experiment through observation was what was needed.

Sadly, Vesalius did not enjoy a long and happy life. He fell out with a patron and embarked on a pilgrimage to the Holy Land. (Some say his conscience got the better of him and he went to atone for all the bodies he had desecrated, though it is extremely unlikely that he ever had such qualms.) On his way home in 1564 he was shipwrecked on a Greek island where he starved to death. He was just 50 years old.

THE BODY AS MACHINE

The University of Padua, where Vesalius did much of his pioneering work, continued to build on its reputation as a leading and forward-thinking medical centre. In 1597, 54 years after *De Fabrica* was published, a young doctor from England called William Harvey went to study there. Harvey was the son of a farmer and had studied medicine at Cambridge, but found the teaching dull and uninspiring and so moved to Padua. There he started to develop ideas that were every bit as unsettling as those of Vesalius. The irony is that William Harvey was not looking to change the status quo – he was a traditionalist at heart. He took a long time to publish his findings, partly because he feared ridicule and partly because he worried about how they would be interpreted. Like Copernicus, he was a reluctant revolutionary, forced into a situation in which he could not help but destroy one of the few elements of Galen's work that remained unquestioned. In doing so, he started a movement that would firmly establish a mechanistic view of the body, something that as a God-fearing traditionalist he would have hated.

When Harvey arrived in Padua he studied under Girolamo Fabrizio, also known as Fabricius. Fabricius, a surgeon, is now chiefly famous for being the first to describe the valves inside veins. Although he created detailed drawings of the valves, he was completely mistaken about their purpose. He thought that the valves were there to control the rate at which blood flows from the liver to the other organs. In this he was clinging to yet another of Galen's mistakes. Harvey would later demonstrate the real role of the valves and, by showing that blood circulates through arteries and veins, fatally undermine the Galen world view.

Below: The University of Padua was a great centre for medical and scientific research between the 15th and 18th centuries.

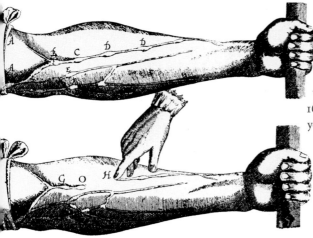

Above:
Illustrations from Harvey's book demonstrate the direction of blood flow through the veins of a human arm.

In Harvey's time, blood itself was seen as one of the four humours – the four basic substances that were thought to fill the human body. According to 16th-century medicine, these substances were phlegm, yellow bile, black bile, and blood. The critical thing was to have a balance of these humours; too much blood, for instance, could cause sickness. And if you believed that then bloodletting, an extremely popular medical practice as late as the 19th century, was a completely logical treatment. But the way that doctors viewed blood – what it is and what it does – was still largely dependent on ideas that were extremely ancient. It was widely accepted, and consolidated by Galen's teachings, that humans have two different blood systems, arteries and veins, which are not connected and which have completely different roles in nourishing the body. According to Galen, the liver made blood, which travelled to the rest of the body via the veins and was totally consumed in the course of its journey. The arteries, on the other hand, carried something called "vital spirits" from the lungs to body, this being their primary task. This was clearly wrong, but not a bad guess considering that oxygen would not be discovered until Priestley and Lavoisier's experiments in the late 18th century (see Chapter Two).

ISLAMIC MEDICINE

In medieval times, the practice of medicine was vastly more sophisticated in the Islamic world than in the West. Islamic scholars collected and then translated into Arabic huge amounts of written work that had been passed down from both the Ancient Greeks and from the Indians. They not only made these writings more accessible and comprehensible, but also built on what they contained. One of the greatest physicians of the Islamic world was a man now known as Avicenna, who lived during the 11th century. Among other things, he realized that diseases like leprosy are contagious and that they can be spread by close contact, and emphasized the importance of quarantine for controlling the outbreaks of plague. His books were widely used in European universities well into the 17th century. Muslim doctors created what might be described as the first modern hospitals, where the sick were looked after by properly trained physicians. They were clean and organized; by contrast, hospitals in Christian Europe were disorganized, filthy, and provided few effective treatments beyond bed rest. Thanks to the work of Avicenna, Islamic hospitals were created with separate wards, so that patients who had contagious diseases could be kept apart from other patients.

THE EXPERIMENTALISTS

Returning to England in 1602, Harvey's society connections enabled him to become a member of an elite club, the Royal College of Physicians. In his new role as a Fellow of the College, he was expected to give a three-day lecture on physiology for its esteemed members. Since it was difficult – or rather, impossible – to demonstrate the physiological processes of living bodies in corpses, Harvey had to be inventive. He introduced animal vivisection into his lectures, and demonstrated what he safely could on humans. His search for suitable demonstrations led him to explore the relationship of the lungs and heart, which focused his attention on the heart and, in turn, on blood.

Harvey believed, like almost everyone at the time, that blood was created continuously in the liver. But how could he demonstrate the hepatic production of blood to the members of the Royal College? Slowly, he came to an awful realization: he could not, because everything he tried showed clearly that what Galen had claimed was utterly wrong. Under close scrutiny, Galen's description of the way that blood is created was impossible to accept. Harvey had started simply enough by measuring the capacity of the heart and working out how much blood it must be pumping per minute, but to his horror, even using the most conservative estimates, he had arrived at a figure of well over 240kg (530lb) of blood a day – more than three times the entire body weight of a man. This made no sense at all. He worked and reworked his calculations, but the numbers were always the same. He reluctantly concluded that Galen must be wrong and that the body could not possibly be making so much blood and destroying it in such a short time scale.

Further experiments soon convinced Harvey that there was a far more plausible explanation for what was going on inside the body. Arteries and veins must be connected and forming a circulating system. Unfortunately, microscopes powerful enough to pick out capillaries, the small vessels that link the arteries and veins, did not yet exist, so Harvey had to try and prove his theory by indirect methods. Some of his most famous sets of experiments were those he performed on himself. First, he tied ligatures around his upper arm until he had cut off all the blood flow to his fingers. Then he slowly loosened the ligatures until the veins in his lower arm (below the ligature) began to bulge. He explained that this was happening because when he loosened the ligatures, blood from the arteries could now reach his fingers. But this blood could not drain through the veins back to the heart because it was still being blocked by the ligature; veins, as we now know, are shallower than arteries and need

Below: The flow of blood between organs according to Galen, with blood moving through the body in tides rather than circulating.

Right: Illustration of an early attempt at a blood transfusion, replenishing a person's blood supply with that from a lamb.

Left: William Harvey 1578–1657. Harvey's new approach to medicine helped to establish a new paradigm of experiment and observation rather than reliance on old authorities.

EXERCITATIO,
ANATOMICA DE
MOTV CORDIS ET SAN-
GVINIS IN ANIMALI-
BVS,
GVILIELMI HARVEI ANGLI,
Medici Regii, & Professoris Anatomiæ in Col-
legio Medicorum Londinensi.

FRANCOFVRTI,
Sumptibus GVILIELMI FITZERI.
ANNO M. DC. XXVIII.

Left: Frontispiece of Harvey's revolutionary book. While his findings were controversial, they did little to affect his reputation as a physician.

less pressure to block them. He also deduced the real purpose of the valves that his former tutor had discovered: they are there to ensure that blood flows – and particularly when it has to do so against the force of gravity – in one direction, back to the heart.

As a result of these and other experiments, Harvey was soon convinced that blood must circulate round the body, driven by the heart. Strong in his religious and Aristotelian beliefs, Harvey claimed, "The concept of a circuit of the blood does not destroy, but rather advances traditional medicine." He was completely wrong – his discovery utterly undermined traditional teaching. It would help launch a new way of seeing the body, not as a balance of vital forces but as a complex mechanical machine. Rightly nervous about how his findings would go down, however, Harvey delayed publishing, and instead spent his time developing his medical practice. He became personal physician to James I, who died in 1625, and later to Charles I. But he could not leave his new ideas entirely alone. He was, for example, thrilled when he was able to demonstrate to sceptical colleagues that the heart really is a pump, thanks to an unfortunate accident that had happened to the son of a royal connection, Viscount Montgomery. The Viscount's son had fallen from a horse when he was a boy, leaving a gaping hole in his chest. As Harvey wrote, "It was possible to feel and see the heart's beating through the scar tissue at the base of the hole."

He finally published in 1628, more than 12 years after his first experiments. The book was called *Exercitatio Anatomica de Motu Cordis et Sanguinis in Animalibus* (*An Anatomical Exercise on the Motion of the Heart and Blood in Animals*). It was received slightly better than he had feared and his reputation survived largely intact. Harvey had unwittingly found the first evidence for a more mechanistic view of the body, but although his breakthrough answered one question, it raised many more. What does the liver do? And what is the role of the lungs? Accepting that blood circulates meant abandoning the security and completeness of accepted wisdom.

When England erupted into Civil War, Harvey remained with the King as his personal physician, eventually retreating to the Royalist stronghold of Oxford. Here, he inspired a group of young experimentalists, including Robert Boyle, who would later become known for his observations on the pressure of trapped air (see Chapter Two), and Robert Hooke. While at Oxford, they researched blood transfusions, conception, embryology, and a wide range of other disciplines. Harvey immersed his disciples in the concept of "experimental science", and started a tradition in England that would shape scientific investigation for centuries to come. The experimental method was given a further boost when Boyle and others founded the Royal Society in 1662.

Below: An illustration of one of William Harvey's animal dissections during a lecture on the circulation of the blood.

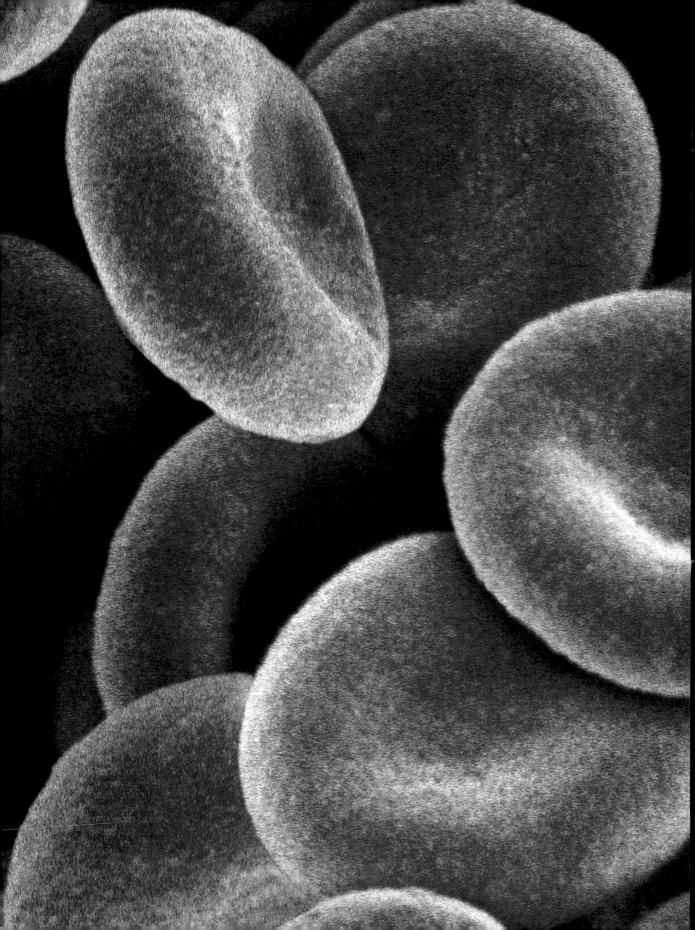

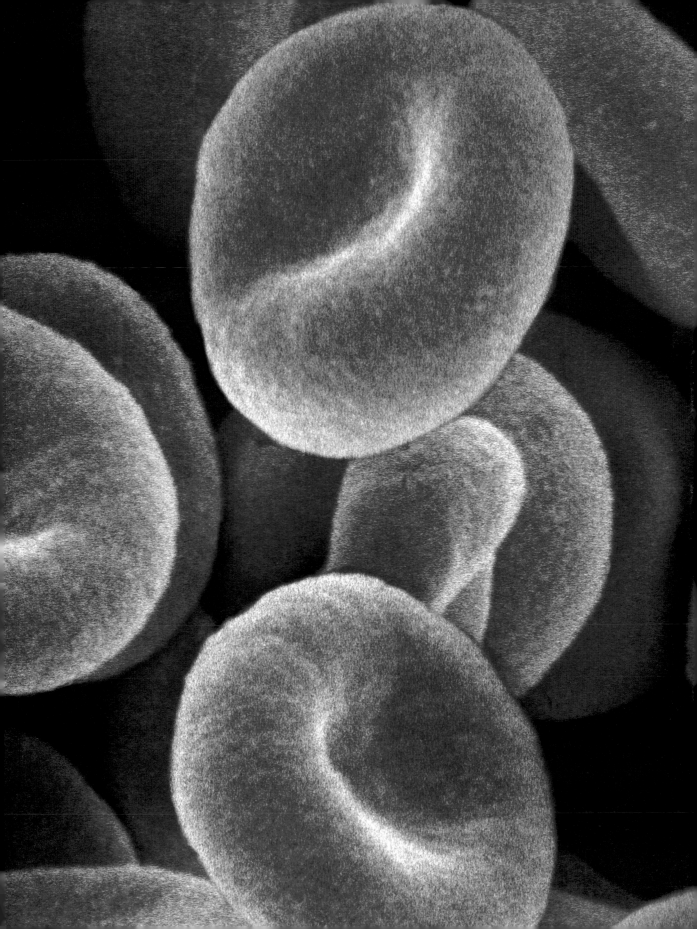

BIRTH OF CHEMISTRY

In the early 1800s, Wilhelm von Humboldt became Education Minister in Prussia and utterly transformed what was seen as the purpose of education, particularly at the university level. As a result of his sweeping reforms, universities became places of research, in which the students participated, rather than simply places where students came to learn by rote. This became the educational model on which the great American universities, such as Yale and Harvard, were founded. The emphasis was now on knowledge as a process rather than as a product. "Education" was seen as an important way of creating considerate citizens, inculcated with intellectual virtues such as self-reliance, autonomy of judgement, and critical thinking.

Humboldt's educational reforms brought with them better funding for universities and created a surge in interest in many different areas of research. Germany, for the first time, overtook Great Britain and France as the country where new technologies would be developed and discoveries made. Students in their droves began to study chemistry, travelling to other countries, learning their secrets and improving on them. This opened up another path to exploring what goes on inside a human life – a path that would be fully exploited by the Germans. Could the recently established discipline of chemistry open up the secrets of life any more effectively than electrophysics had?

An early believer that life can be explained by chemistry was Justus von Liebig. He established a research centre at Giessen in 1824 and pursued bodily functions with an alarming zeal. He made his students analyse, test, and count the constituent chemicals of everything that went into the human body – including, food, gases, and water – and, unfortunately for the students, everything that came out. He showed that the weight of the ingesta (or inputs) exactly equaled the weight of the excreta (or outputs). For him, this was evidence that life can be reduced to a series of chemical equations. Life, he decided, is not that special. Despite his meticulous analytical methods, however, Liebig's approach was criticized. The great French physiologist, Claude Bernard, said that it was like trying to work out "what occurred in a house by measuring who went in the door and what came out of the chimney". So, in the search to find the secret of life, the focus was narrowed from looking at the chemical processes of the whole body to looking for specific chemical reactions occurring inside the body.

Left: Justus von Liebig 1803–1873. Liebig's investigations of the human body led him to develop important chemical processes – his inventions included artificial fertilizers and the Oxo cube.

Right: An illustration of Liebig's workbench shows the curiously shaped "Liebig condenser", still widely used in school laboratories.

By the middle of the 19th century, the spread of "germ theory" and the growing acceptance that microbes can kill us had once more focused people's attention on a world only visible through a microscope. Important things seemed to happen at the cellular level, but what was actually going on inside the cell itself?

Friedrich Miescher
1844–1895

In 1868, Dr Friedrich Miescher arrived at a castle in Tübingen in Germany to study blood. For his research he needed copious amounts of pus, a substance rich in infection-fighting white blood cells. This particular part of Germany had been at war with Prussia and there were plenty of injured soldiers lying around with weeping wounds. Miescher took pus-soaked bandages and used an enzyme called pepsin, which he got by scraping out the mucus that lines the stomachs of pigs, to break down the cells. Once the pepsin had destroyed the walls of the white cells, he was able to study the contents of the nucleus. He found, as he expected, that it contained lots of carbon, hydrogen, nitrogen, and oxygen, but he also showed that it contained phosphorus. This was surprising because he had imagined that the substance at the heart of the cell would be a protein, and proteins do not contain phosphorus. Whatever it was that he had found in the nucleus was something new. He called it nuclein, because it came from the nucleus; we know it now as deoxyribonucleic acid or DNA. Curious, Miescher began looking for nuclein in other human cells and also in the cells of a wide range of different creatures, from frogs to salmon. And wherever he looked, he found it. DNA was clearly universal and important, though Miescher himself, it seems, never realized just how important. Nor, for nearly 60 years, did anyone else.

"Miescher began looking for nuclein in other human cells and also in the cells of a wide range of different creatures. And wherever he looked, he found it."

Right: An artist's impression of the structure of Miescher's "nuclein" – the famous double-helix of DNA.

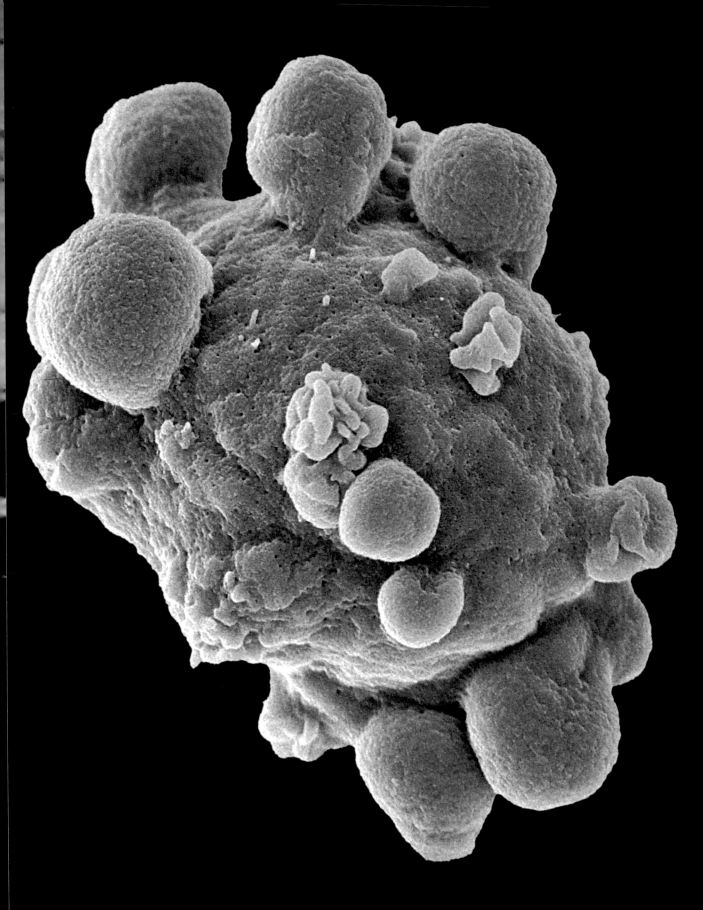

The discovery of the structure of DNA was a great moment in the history of science and in the search to discover the secret of life; understanding the structure answered some hugely important questions about replication, mutation, and human evolution. Since 1953 there have been further tremendous advances in understanding exactly how DNA replicates, how it makes proteins, how it can go wrong, and how it is that such a small number of genes (only about 30,000) can produce something as complicated as a human being. In the light of what we have learnt in just half a century it is tempting to claim, as some enthusiasts do, that we are on the brink of being able to truly control and manipulate life. The problem is that, as scientists discovered in the past, the closer we look the more complicated it gets. But although we may not yet know exactly what it is that makes us tick, we do know that one extraordinary molecule is at the heart of us and every other life form on this planet.

Left. An electron microscope image of a blood cell stricken by leukaemia. The ability to study cells and other bodily structures in such detail can reveal hidden aspects of their function and even point to new treatments for diseases.

MUTATION

Mutation sounds terrible, like the unfortunate product of a dangerous experiment leading to the creation of some freakish or unfortunate monster. Yet if it was not for mutation we would not exist because our remote ancestors would never have emerged from the primeval ooze; favourable mutations play a crucial role in evolution, providing a source of variation on which natural selection acts. A mutation can occur when a cell divides and the letters in a stretch of DNA that forms a particular gene are not faithfully copied. A stretch of DNA reading AACCCG, for instance, could be copied as AGCCCG. When this stretch of DNA is later used to make a protein, that protein will be different to proteins produced from the original sequence. This may or may not be important. It may lead to the death of the animal, or just a subtle change, or no change at all. As a simple example, imagine a mutation changes the colour of a moth's wings from light to dark. If the moth lives on light-coloured trees then this mutation will make the moth stand out. Moth and mutation will rapidly become extinct. But if the trees become blackened – perhaps because of soot from a nearby factory – then the dark moths will blend better into this new environment. The mutation is favoured. We will see far more dark-coloured moths. Mutation has led to evolution.

Connections – Body

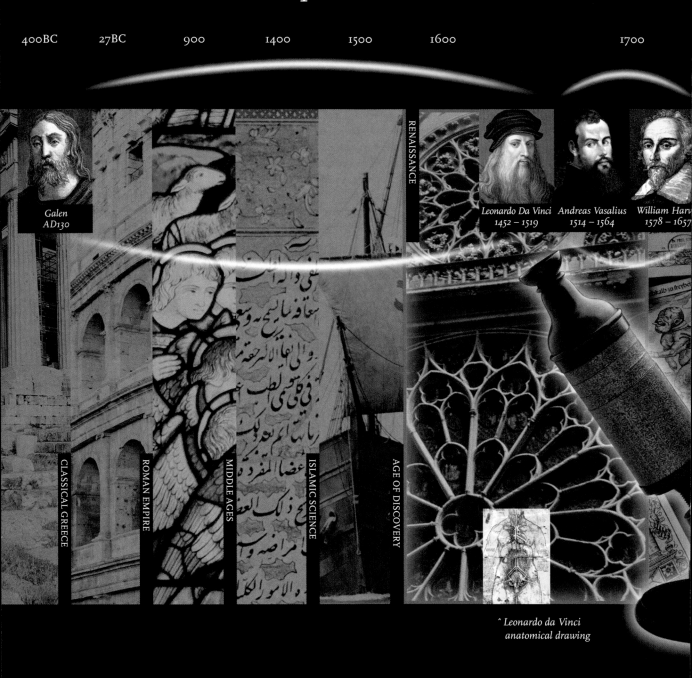

400BC   27BC   900   1400   1500   1600   1700

Galen
AD130

RENAISSANCE

Leonardo Da Vinci
1452 – 1519

Andreas Vasalius
1514 – 1564

William Harv
1578 – 1657

CLASSICAL GREECE

ROMAN EMPIRE

MIDDLE AGES

ISLAMIC SCIENCE

AGE OF DISCOVERY

^ Leonardo da Vinci
anatomical drawing

Our attempts to understand and prolong life have drawn us into ever more detailed inspections of the workings of the body. Unfortunately, the view of our anatomy that held sway until the Renaissance was based on dissections of animals by a Greek called Galen. The Church discouraged dissecting bodies, stifling further research. Leonardo da Vinci's incredibly accurate drawings were never seen widely; it was only when Andreas Vesalius, a professor of anatomy at Padua, bravely started an unprecedented programme of dissections that a reasonably accurate anatomical guide was created. Fifty-four years later a young British doctor, William Harvey, discovered the circulation of the blood, further discrediting Galen's work.

1800   1900   2000

Antonie van Leuuwenhoek 1632 – 1723

Robert Hooke 1635 – 1703

Rosalind Franklin 1920 – 1958

James Watson 1928 –

Francis Crick 1916 – 2004

AGE OF ENLIGHTENMENT

EARLY 20TH CENTURY

MID 20TH CENTURY

21ST CENTURY

< Robert Hookes' microscope

^ DNA double helix structure
^ DNA x-ray

Our understanding of the body – and how to protect it – took on a new dimension in the 17th century with the use of microscopes. Robert Hooke was able to observe the cell, the building block of life itself; Antonie van Leeuwenhoek's far more powerful microscope was able to see microbes, "animalcules", paving the way for germ theory and the work of scientists like Louis Pasteur in preventing infectious disease.

From the late 19th century our focus zoomed in further still, to what made cells work. By the mid 20th century, it was clear that genes were crucial and that they were made of mysterious substance, DNA. Now, using x-rays rather than telescopes, Rosalind Franklin was able to photograph individual strands of DNA, from which Crick and Watson defined its structure, answering many questions about the body, but posing still more.

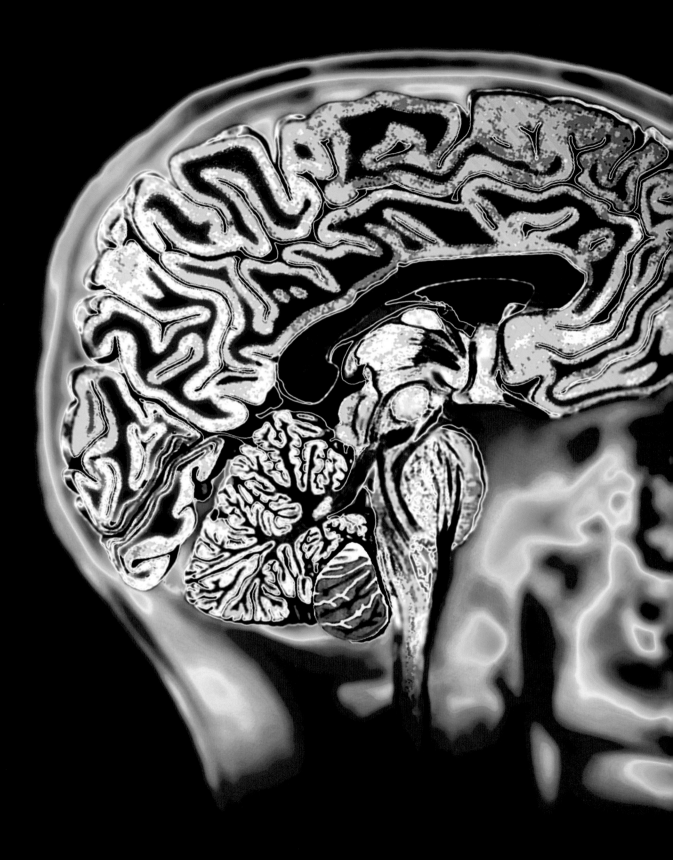

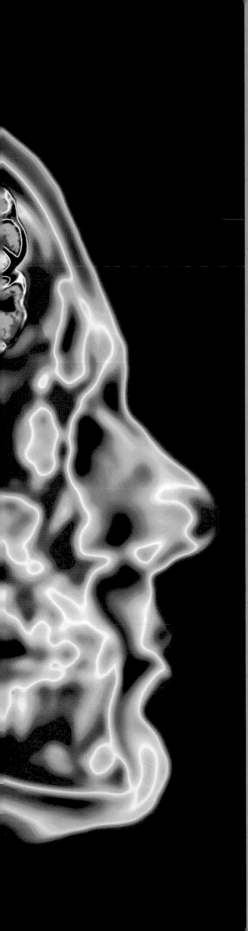

Mind

WHO ARE WE?

If you were asked to describe yourself, you would probably start with your physical appearance. But when it comes to describing precisely what it is that drives and motivates us, most of us would struggle. The truth is that few of us really understand the workings of our own mind; we often behave irrationally, procrastinate, and do unpredictable things for reasons that are not obvious even to ourselves. The search to understand who we are and what really motivates us has been a long one. Early human civilizations had little idea that the brain was responsible for cognition at all, and it has only been in the last century that we have begun to understand how the 1.5kg (3lb) of grey and white matter that sits on top of our spinal cord allows us to think. But neuroscientists are now unearthing increasing evidence that much of what the brain does lies beneath our conscious awareness and that many of the decisions we think we make are actually rationalizations – the conscious part of us justifying decisions that have, in fact, already been taken by the subconscious parts.

Our brains, like our bodies, are the product of an extremely long process of evolution. On top of a 500 million-year-old reptile brain, containing the brain stem and cerebellum, is grafted a much younger addition: the area of the brain called the neocortex, which is responsible for language and abstract thought. The neocortex, Latin for "new bark", is a thin layer of grey and white matter that sits on top of the cerebral hemispheres – the left and right sides of the brain. It is the seat of rational thought, and is seen by some as a sort of "command centre". But the neocortex is actually more like the mahout who perches on top of the elephant; sometimes he can direct the elephant along the path that he wants the elephant to take, but often the elephant goes where it wants to go and the mahout is left pretending that, yes, that is also where he wanted to go.

Left: A magnetic resonance imaging (MRI) scan of a human skull reveals the complex folds and details of the brain – the body's most complex and intriguing organ.

THE BRAIN COMES TO LIFE

When asked who they really are, most people will try to describe their behavioural characteristics or way of thinking – today, we are beginning to see our true selves as the workings of our mind. But the first people to seriously consider who they were in secular as well as religious terms were the classical Greeks. Hippocrates, a famous physician, not only worked out that injuries to the brain affected opposite sides of the body but also understood that epilepsy was a brain disease, not possession by an evil spirit. And while the Greeks believed that their destinies were determined by mythical beings called "Fates", they thought that day to day behaviour was controlled by an individual's psyche.

Hippocrates
c.460–c.377 BC

Plato, who lived in the 4th and 5th centuries BC, was one of the first to develop the idea that the self, responsible for our rational decisions, our reason, and our thoughts, was associated with the brain. Plato's theory, based more on philosophy than clinical examination, was that our cognitive soul, which would pass to another body after death, was made up of extremely fast-moving spherical particles, which concentrated in the head and nervous system. He suggested that other organs had psyches too, although these were not immortal. One psyche, associated with desire or animalistic appetites, inhabited the diaphragm from where it could easily energize the nearby liver, believed to be the organ of lust. Another – the psyche that controlled emotions – was seated in the heart. (The use of hearts on Valentine's cards is a throwback to these ideas.)

Right: Plato was famous for his mathematical and philosophical theories. His ideas about the mind were derived from idealized precepts rather than careful observation.

Aristotle, who had first learnt about the human body from his father, a personal doctor to the King of Macedonia, jumped on the bandwagon of physiological speculation. Taking a great leap backwards, he taught the civilized world that the heart was the seat of the rational soul, while preaching that the brain was little more than a bloodless radiator that allowed the body to cool off. Strokes, in his opinion, were blockages in the head caused by a build-up of black bile. However, his idea of lasting value was that there was a mysterious, almost heavenly, substance that the centralized psyche could send to the muscles to make them function. Of course, he had no knowledge of electrical impulses being conveyed along neurons or of chemical transmitters crossing synapses, but his concept pointed future generations in the right direction. Aristotle thought that all empty spaces contained a new and powerful – but invisible – fifth element called aether. This was taken into the body by the lungs and was then transformed into "vital pneuma" or "vital heat" in the heart, the organ of perception from which he supposed all the nerves emanated. The vital pneuma, a life- giving force, was then carried by the blood into muscles where it would activate their psyches, spurring them into action. Thus, what Aristotle understood was that there was one organ controlling all the others. He simply had not worked out that it was the brain.

Below: A medieval Italian relief showing Plato and Aristotle – two great philosophers whose ideas on the mind proved sadly mistaken.

PROBING THE DEAD... AND THE LIVING

One reason that the classical Greeks failed to advance neurology beyond their ingenious but speculative philosophies was that they had beliefs about treatment of the dead that severely restricted what they might have learned by carrying out dissections. They thought that if a corpse was not properly attended to – in a quick and respectful burial – a person could spend eternity wandering the bleak banks of the River Styx. For this reason, human autopsies were strictly illegal. Even animal dissections were not encouraged as many classical Greeks believed in reincarnation. But with the arrival of the Hellenistic Age, heralded by Alexander the Great and the founding of his capital in Egypt rather than on the Greek mainland, religious ideas began to change. The soul and the body were no longer seen as being so intimately connected and pathologists were able to sharpen their scalpels.

Herophilus of Chalcedon, a Turkish Greek living in Alexandria, performed hundreds of human dissections and in doing so made the first detailed examinations of our brains. Slicing them into sections, he noted the four hollows (ventricles) within them, separated the two membranes that cover them, described their two major parts (cerebellum and cerebrum), and traced what he recognized as motor or sensory nerves extending from the brain stem to other regions of the body. While he made some interesting discoveries, however, his methods were exceptionally brutal. Ptolemy, the King of Egypt at the time, was so keen to explore the human brain that he presented Herophilus with condemned criminals on whom he could perform his horrific vivisections.

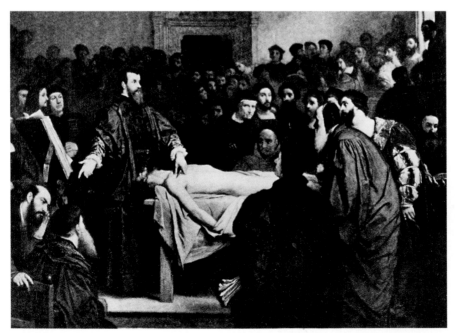

Left: Herophilus conducts a dissection before watching students in his medical school at Alexandria.

Above right: A medieval illustration of Herophilus and Erasistratus, his apprentice at Alexandria who continued the early tradition of studying anatomy through dissection.

IOCLES §HEROPHIL9§ERASISTRAT9§

TREPANNING AND EARLIEST BRAIN SURGERY

Although the study of the brain is still in relative infancy surgeons have been successfully opening the cranium for over 7,000 years. Trepanning, trephining, or craniotomy is a practice that involves cutting out a section of skull and letting it heal over. It was first used in Neolothic Europe well before the creation of Stonehenge. Skulls discovered at Ensisheim in France display neat apertures, often several inches across, opened up with flint blades long before their owners' deaths. Other early cultures, stretching from China to Central America, were also avid trepanners. Although not strictly brain surgery – piercing the organ's outermost layer, the dura mater, would almost certainly have resulted in death from infection – trepanning could save victims of a head wound suffering from build-up of fluid. By the 15th century, itinerant barber-doctors roamed Europe, opening the skulls of fee-paying patients after claiming they could cure their depression or epilepsy. Some even spread the myth of an extractable "stone of madness", presumably palming a pebble after the trepanation. Even today, a more sophisticated form of the practice is not only used by surgeons to relieve swellings beneath the skull but also attracts a cult following among a self-practicing New Age community who believe it can enhance well-being.

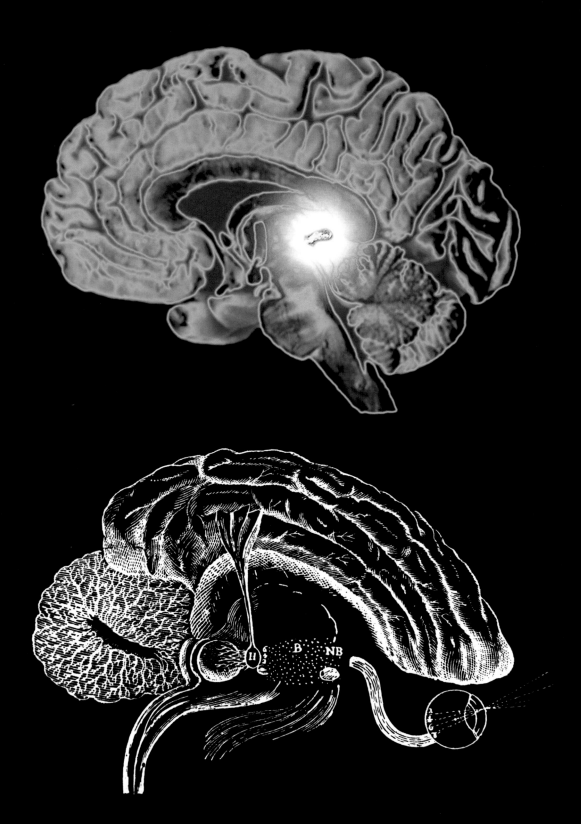

Descartes wrote that the gland was able to swivel and move around, distributing special animal spirit particles along the required nerves, which contained trap doors attached by fibres that could open or shut. He said that these particles flowed out of the ends of nerves to make the muscles swell, physically forcing them to expand while the opposite muscles contracted. According to Descarte's theory, the mind was the conscious pilot controlling the body, with the ability to override its usual actions. Subconscious activities were carried out by reactionary reflexes. Physical sensations tightened or slackened fibres in the nerves causing muscles to relax or contract. And while it had been believed that the soul provided the body with its vital heat, now a model was in place that suggested the body was independent and could survive without the soul, explaining the continuing existence of the unreasoning and ungodly kingdom of beasts. He decided that animals did not have souls and were little more than sophisticated mechanical machines.

Even though Descartes' separation of the soul from the body should have been appreciated by the Church – it removed the thorny issue of pigs in heaven – his theories were regarded as scandalous. He spent the rest of his life on the run from the dreaded Inquisition. But while his speculation about human mechanics never gained much support (anatomists knew that animals also had pineal glands), his legacy is its philosophical accompaniment. Descartes reasoned "*cogito ergo sum*" or "I think therefore I am", suggesting that the power of our consciousness and cognition makes us who we are. After the shock of discovering that the Earth was not at the centre of the Solar System, his theories gave us reason again to believe in our own importance, as the only creatures with minds. We were not an inconsequential nothing but the centre of our own Universe. The workings of the brain became the key to the secret of what makes us human.

Left: A modern computer-generated cross-section of the human brain (top) with Descartes' all-important pineal gland highlighted, compared with Descartes own interpretation (the pineal gland is marked with an "H").

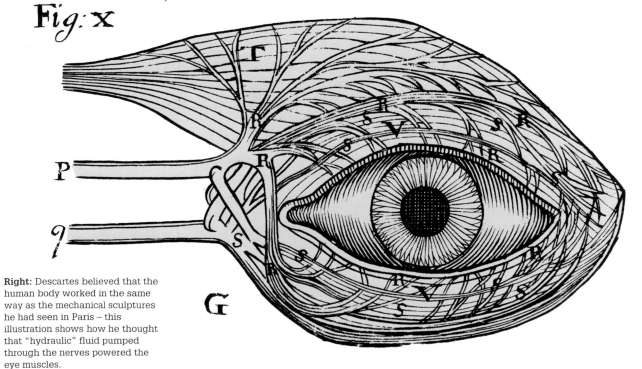

Right: Descartes believed that the human body worked in the same way as the mechanical sculptures he had seen in Paris – this illustration shows how he thought that "hydraulic" fluid pumped through the nerves powered the eye muscles.

islanders who have been isolated for tens of thousands of years, shows that we did not learn these characteristics. They must have been inherited from our common ancestors. He realized that we feel and display similar emotions as animals.

The claim that human passions are simply a continuum of animal survival patterns seemed to fit with a theory first suggested by Aristotle: we are like animals in every way apart from a higher soul – now seen as our reasoning or mind – that is able to suppress or override our base desires. This idea was very popular in the stiff and tightly buttoned Victorian age. But Darwin's theory also opened a Pandora's box. It was now impossible to scientifically consider who we are without delving deeper, going beyond our purely logical, rational, and conscious minds.

"Darwin's theory also opened a Pandora's box. It was now impossible to scientifically consider who we are without delving deeper, going beyond our purely logical, rational, and conscious minds."

Right: Another page of striking plates from Darwin's book on emotions. More than 5,000 copies were sold, making the book something of a bestseller despite its luxurious and expensive production.

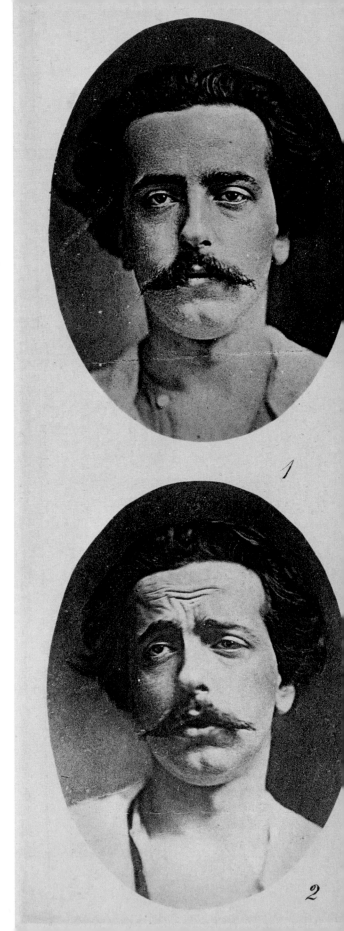

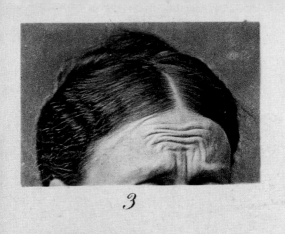

3

6

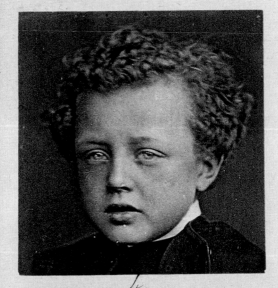

4

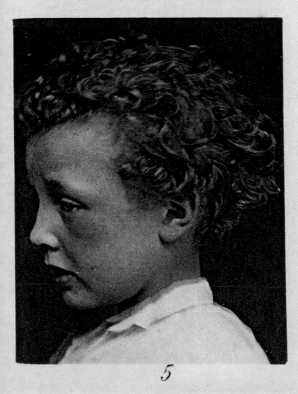

5

7

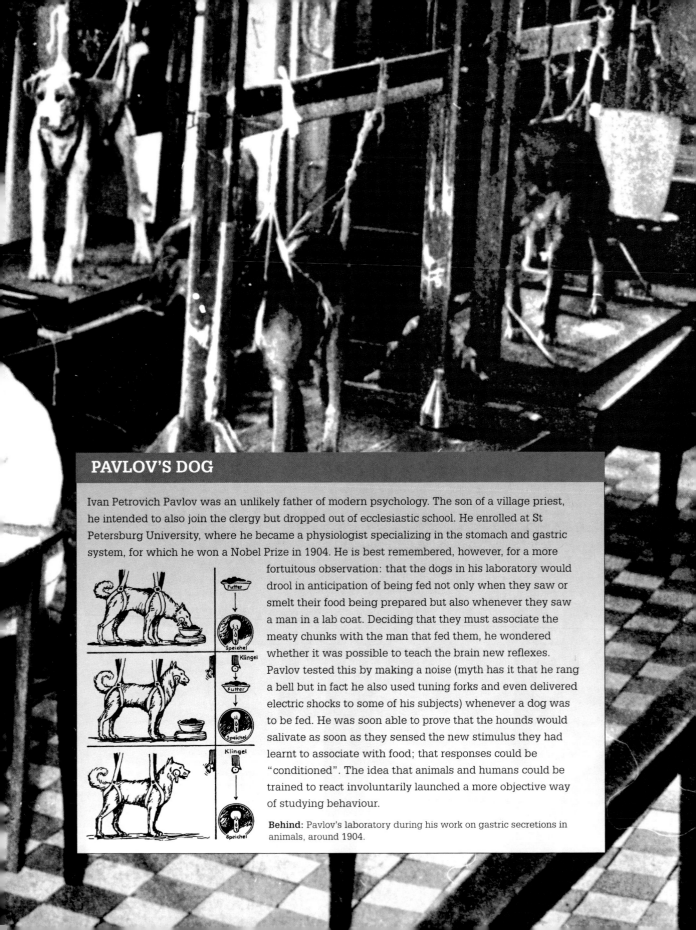

PAVLOV'S DOG

Ivan Petrovich Pavlov was an unlikely father of modern psychology. The son of a village priest, he intended to also join the clergy but dropped out of ecclesiastic school. He enrolled at St Petersburg University, where he became a physiologist specializing in the stomach and gastric system, for which he won a Nobel Prize in 1904. He is best remembered, however, for a more fortuitous observation: that the dogs in his laboratory would drool in anticipation of being fed not only when they saw or smelt their food being prepared but also whenever they saw a man in a lab coat. Deciding that they must associate the meaty chunks with the man that fed them, he wondered whether it was possible to teach the brain new reflexes. Pavlov tested this by making a noise (myth has it that he rang a bell but in fact he also used tuning forks and even delivered electric shocks to some of his subjects) whenever a dog was to be fed. He was soon able to prove that the hounds would salivate as soon as they sensed the new stimulus they had learnt to associate with food; that responses could be "conditioned". The idea that animals and humans could be trained to react involuntarily launched a more objective way of studying behaviour.

Behind: Pavlov's laboratory during his work on gastric secretions in animals, around 1904.

AND BEYOND...

Just as the telescope, in the 17th century, extended human understanding of the Universe, so modern brain-scanning techniques, like functional magnetic resonance imaging (fMRI), are opening the door to a new wealth of information about the workings of the brain. We now know, for example, that we have neurons so specific that they appear only to be interested in one thing. Scientists have found neurons that only become active when a person thinks about the actress Jennifer Aniston; Bill Clinton and Halle Berry also have their own neurons. And as early as 1960 a neuron was identified that is only activated when a person thinks about their grandmother. Brain scanners can also reveal the areas of the brain that are active when a simple choice is being made, such as which button to press out of a choice of two. This has allowed experimenters to predict their subject's decision several seconds before the person being analysed is even aware that they have made that choice.

While we have become increasingly aware of how complex our brains are, however, we have also become aware that our brains, like ourselves, are prone to taking short cuts. Most people think of vision, for example, as being like a video camera, faithfully recording what is out there. In fact it is nothing like that. The brain tends to create its own reality and ignore most of what it sees. In one test, psychology students on their way to an interview met a receptionist sitting behind a desk. They were then briefly distracted by someone shouting their name, during which time the receptionist swapped with someone completely different. When the students turned back to the receptionist, none of them noticed they were now talking to someone completely different. Scientists searching for the cause of road accidents have discovered even more disturbing implications of this corner cutting. If you are driving down an empty street and glance away, when you look back the brain assumes it is looking at much the same image it was looking at a second before. There is often a significant time delay before it registers that the scene has changed and that there is now someone standing in the middle of the road. This phenomenon is known as change blindness. The reason it happens is that the extraordinary quantity of information that we are constantly receiving through our senses cannot all be processed instantaneously. The brain is forced to take short cuts, to make assumptions.

"Modern brain-scanning techniques, like functional magnetic resonance imaging (fMRI), are opening the door to a new wealth of information about the workings of the brain."

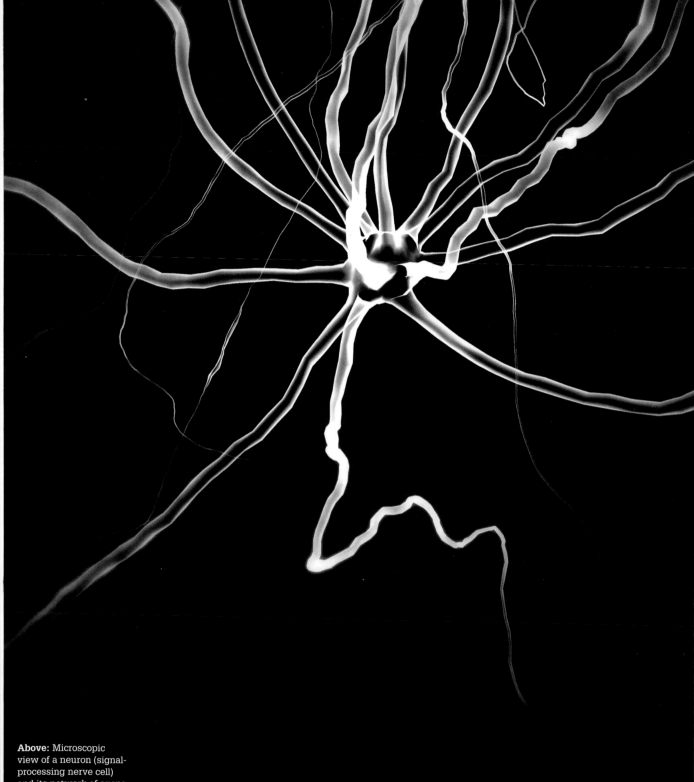

Above: Microscopic view of a neuron (signal-processing nerve cell) and its network of axons or nerve fibres – long extensions used for transmission of signals between neurons.

Connections – Mind

Hippocrates c460 – c370 BC

Plato c428 – c348 BC

Aristotle c384BC – c322 BC

ISLAMIC SCIENCE

AGE OF DISCOVERY

René Descartes 1596 – 1650

Thomas Willis 1621 – 1675

RENAISSANCE

REFORMATION

CLASSICAL GREECE

ROMAN EMPIRE

MIDDLE AGES

^ *Trepanning*   ^ *Egyptian manuscript describing effects of brain injuries*

We have been trying to understand human identity and motivations for thousands of years. Agreement that this search should focus on the brain is surprisingly recent, however. In ancient Greece, Hippocrates, like the early Eygptians before him, deduced from the effects of injuries to the brain that it had some influence on controlling the body, a belief possibly reflected in the practice of trepanning. Plato built further on his idea, but Aristotle relocated the rational soul to the heart, where it stayed for a thousand years – adopted and policed by the Church.

The anatomical observations of Andreas Vesalius and later René Descartes put the search back on track, the latter developing the theory of dualism – that the mind was separate to the body, "piloting" it from a cockpit in the pineal

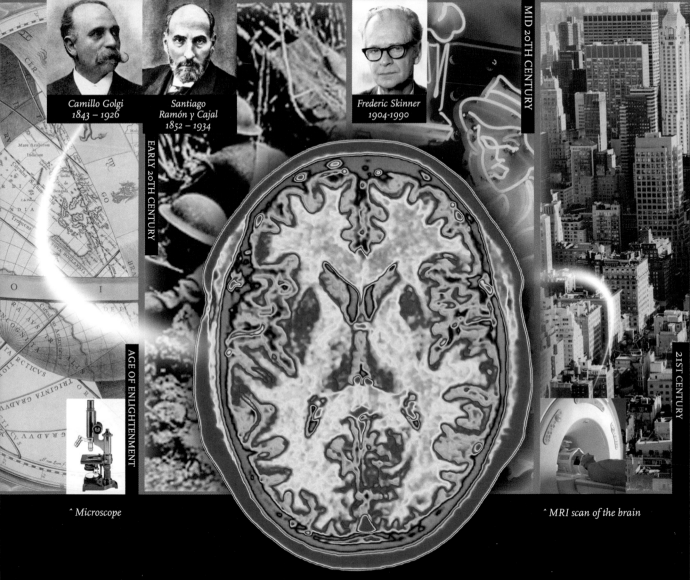

Camillo Golgi
1843 – 1926

Santiago
Ramón y Cajal
1852 – 1934

Frederic Skinner
1904-1990

MID 20TH CENTURY

EARLY 20TH CENTURY

AGE OF ENLIGHTENMENT

21ST CENTURY

^ Microscope

^ MRI scan of the brain

gland in the brain. Shortly after, Thomas Willis became the first to systematically examine the anatomy of the brain, and to definitively state that it determined who we are.

The search for the self was far from over though. Investigations continued – into how our minds differed from other creatures, into how we could be conditioned and controlled (famously by Pavlov and Skinner), and into how we could be cured of mental diseases. Throughout, the ever-improving ability to understand the physical structure and processes of the brain was crucial – early microscopes enabled Camillo Golgi and Santiago Ramón y Cajal to reveal the detailed structure of the brain and the existence of neurons. Today, with techniques such as MRI scanning, we are even able to see the brain in action.

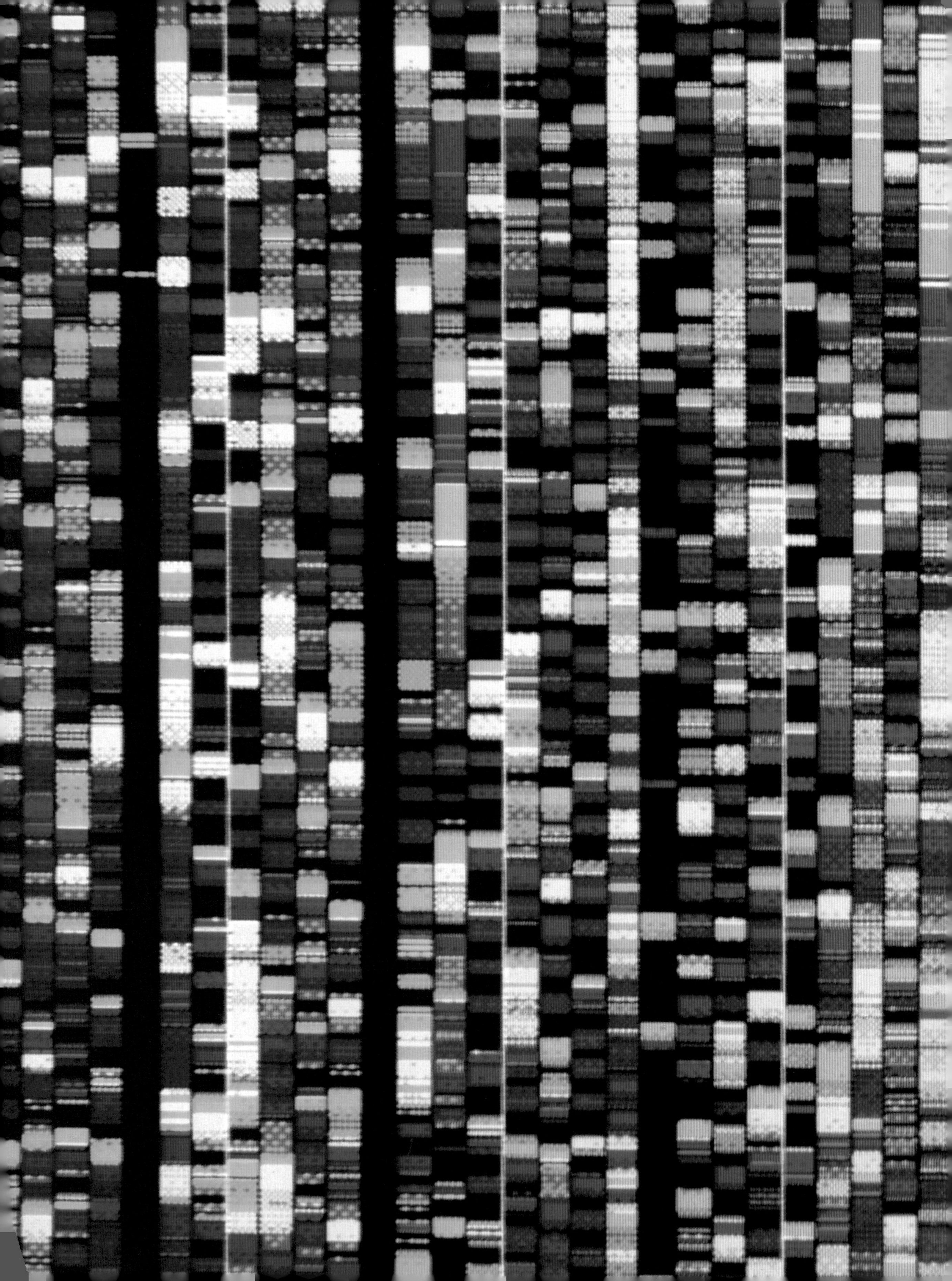

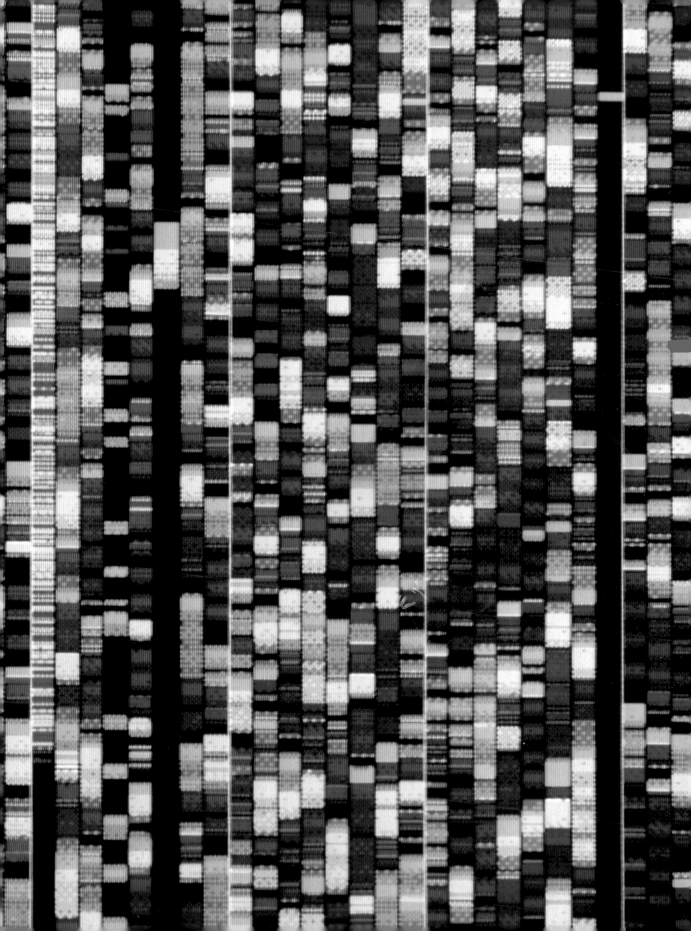

FURTHER READING

General

Bryson, Bill *A Short History of Nearly Everything* (Black Swan 2004)
Dunbar, Robin *The Human Story* (Faber & Faber, 2005)
Dunbar, Robin *The Trouble with Science* (Faber & Faber, 1996)
Roberts, Royston M. *Serendipity: Accidental Discoveries in Science* (John Wiley & Sons, 1989)
Waller, John *Fabulous Science: Fact and Fiction in the History of Scientific Discovery* (Oxford University Press, 2004)
White, Michael *Rivals: Conflict As the Fuel of Science* (Vintage, 2002)
Youngson, Robert *Medical Blunders* (Robinson Publishing, 1996)
Youngson, Robert *Scientific Blunders* (Robinson Publishing, 1998)

Chapter One: Cosmos

Biagioli, Mario *Galileo Courtier* (University of Chicago Press, 1993)
Fara, Patricia *Science: A Four Thousand Year History* (Oxford University Press, 2009)
Gingerich, Owen *The Book Nobody Read* (Heinemann, 2004)
Gleick, James *Isaac Newton* (Pantheon Books, 2003)
Gribbin, John *Science: A History* (Alan Lane, 2002)
Jardine, Lisa *The Curious Life of Robert Hooke* (HarperCollins, 2003)
Sharov, Alexander S. and Novikov, Igor D. *Edwin Hubble, the Discoverer of the Big Bang Universe* (Cambridge University Press, 2005)

Chapter Two: Matter

Bell, Madison Smartt *Lavoisier in the Year One: The Birth of a New Science in an Age of Revolution* (W.W. Norton & Co., 2006)
Fara, Patricia *Pandora's Breeches: Women, Science & Power in the Enlightenment* (Random House, 2004)
Gribbin, John *Deep Simplicity: Bringing Order to Chaos and Complexity* (Penguin Press, 2009)
Holmes, Richard *The Age of Wonder: How the Romantic Generation Discovered the Beauty and Terror of Science* (HarperPress, 2009)
Lavoisier, Antoine *Elements of Chemistry* (Dover Publications, 1984)
McClellan, James E. and Dorn, Harold *Science and Technology in World History: An Introduction* (The Johns Hopkins University Press, 2006)

Chapter Three: Life

Darwin, Charles *The Origin of Species* (Penguin Classics, 1985)
Gould, Stephen Jay *Time's Arrow, Time's Cycle* (Harvard University Press, 1987)
Gribbin, John and Gribbin, Mary *Flower Hunters* (Oxford University Press, 2008)
Holmes, Richard *The Age of Wonder* (Harper, 2008)

McCoy, Roger M. *Ending in Ice: Alfred Wegener's Revolutionary Idea and Tragic Expedition* (Oxford University press, 2006)

Quammen, David *The Reluctant Mr Darwin* (Norton & Co, 2006)

Winchester, Simon *The Map That Changed the World* (Viking, 2001)

Chapter Four: Power

Devreese, J.T. and Berghe, G.Vanden *Magic Is No Magic: The Wonderful World of Simon Stevin* (WIT Press, 2007)

Fara, Patricia *An Entertainment for Angels* (Icon Books, 2002)

Highfield, Roger and Carter, Paul *The Private Lives of Albert Einstein* (St Martins Press, 1994)

Jardine, Lisa *Ingenious Pursuits* (Little Brown, 1999)

Marsden, Ben *Watt's Perfect Engine* (Icon Books, 2002)

Morus, Iwan Rhys *Michael Faraday and the Electrical Century* (Icon Books, 2004)

Chapter Five: Body

Endersby, Jim *A Guinea Pig's History of Biology* (Heinemann, 2007)

Endersby, Jim *Imperial Nature: Joseph Hooker and the Practices of Victorian Science* (University Of Chicago Press, 2008)

Waller, John *The Discovery of the Germ: Twenty Years That Transformed The Way We Think About Disease* (Icon Books Ltd, 2004)

White, Michael *Leonardo: The First Scientist* (Abacus, 2001)

Chapter Six: Mind

Pietro Corsi, *The Enchanted Loom* (Oxford University Press, 1991)

Damasio, Antonio R. *Descartes' Error: Emotion, Reason and the Human Brain* (Vintage, 2006)

Damasio, Antonio R. 'How the Brain Creates the Mind' in *Best of the Brain* (Dana Press, New York, 2007)

Doige, Norman *The Brain That Changes Itself* (Penguin, 2008)

Ekman, Paul *Darwin and Facial Expression: A Century of Research in Review* (Academic Press, New York, 1973)

Finger, Stanley *Origins Of Neuroscience* (Oxford University Press, 1994)

McHenry, Lawrence C. *Garrison's History Of Neurology* (Charles C. Thomas, Illinois, 1969)

Panek, Richard *The Invisible Century: Einstein, Freud, and the Search for Hidden Universes* (Harper Perennial, 2005)

Shorter, Edward *A History of Psychiatry: From the Era of the Asylum to the Age of Prozac* (Jossey Bass, 1998)

Stevens, Leonard A. *Explorers Of The Brain* (Angus & Robertson, London, 1973)

Previous page:
A computer screen shows a sequence forming part of the human genetic code. As each person's genetic code is unique, this sequence is called a DNA fingerprint.

INDEX

Page numbers in bold refer to illustrations

PICTURE CREDITS

Alamy/19th era 2 164; /artpartner-images 58, 101 right; /Peter de Clercq 176; /Coston Stock 42 bottom right; /Stephen Finn 107; /steven gillis hd9 imaging 54 background picture 2, 100 background picture 2, 142 background picture 2, 184 background picture 2, 228 background picture 2, 274 background picture 2; /Angelo Hornak 238; /imagebroker 55 background picture 3, 101 background picture 2, 143 background picture 2, 185 background picture 2, 229 background picture 3, 275 background picture 2; /INTERFOTO 37, 54 bottom right, 61, 132 right, 263; /Lordprice Collection 160 left; /Ilene MacDonald 172; /Mary Evans Picture Library 162, 234 top, 274 top left; /Mathew Monteith 112; /North Wind Picture Archives 30, 54 bottom left, 234 bottom; /Gianni Dagli Orti/The Art Archive 243; /Permian Creations 65; /Phototake, Inc. 181 bottom; /Wolfgang Pölzer 170; /Alex Segre 250 left; /The Art Gallery Collection 23, 54 background picture 3, 100 background picture 3, 142 background picture 3, 184 background picture 3, 228 background picture 3, 274 background picture 3; /The Print Collector 89, 101 bottom left, 120; /Ken Welsh 132 left, 143 bottom right; /WILDLIFE GmbH 239; /World History Archive 41, 115, 128, 143 bottom left, 156 left, 203, 245.

Bridgeman Art Library/Bibliotheque des Arts Decoratifs, Paris, France/Archives Charmet 262, 274 bottom right; /Derby Museum and Art Gallery, UK 62; /Faculty of Medicine, Lyon, France 259; /Natural History Museum, London, UK 108; /Private Collection 106; /Private Collection/Archives Charmet 233, 274 centre.

Corbis 42, 142 bottom right, 158 bottom, 202, 66; /James L. Amos 54 background, 100 background picture 6, 184 background picture 6, 228 background picture 6, 274 background picture 6; /Bettmann 43, 77, 109 left, 109 right, 174 bottom, 185 bottom left, 186, 210 left, 228 bottom left, 236, 267; /Barbara Chase 55 background picture 1, 100 background picture 7, 142 background picture 6, 184 background picture 7, 228 background picture 7, 274 background picture 7; /Dennis Kunkel Microscopy, Inc./Visuals Unlimited 212; /Elio Cio 146 bottom; /Cameron Davidson 55 background picture 4, 101 background picture 3, 143 background picture 3, 185, 229, 275 background picture 3; /Christel Gerstenberg

175; /Hulton-Deutsch Collection 214; /Image Source 273, 275; /Reed Kaestner 54 background picture 1, 100 background picture 1, 142 background picture 1, 184 background picture 1, 228 background picture 1, 274 background picture 1; /David Lees 31, 157; /Micro Discovery 198; /NASA 139; /Alain Nogues/Sygma 182, 185 bottom right; /Gianni Dagli Orti 257; /William Perlman/Star Ledger 130; /John Rensten 85 bottom; /Reuters 54 background picture 4, 100 background picture 4, 142 background picture 4, 184 background picture 4, 228 background picture 4, 274 background picture 4; /Stapleton Collection 22.

Fotolia/iQoncept 72 inset.

Getty Images 159, 207; /Willem Blaeu 25 bottom right; /Cornelia Doerr 148; /Dorling Kindersley 50, 55 centre, 63, 82; /Dave King 60 bottom; /Lambert 86; /NASA 8 bottom; /NASA, ESA, A. Aloisi (STScI/ESA), and The Hubble Heritage (STScI/AURA)-ESA/Hubble Collaboration 20; /NASA, ESA, and The Hubble Heritage Team (STScI/AURA) 16; /NASA/ESA/JPL/Arizona State Univ. 53; /Richard Nebesky 18; /Hans Neleman 272, 275 bottom right; /SSPL via Getty Images 87, 88 top, 88 bottom, 94, 100 bottom, 134, 153, 154, 160 right, 172 bottom, 184 right; /Time & Life Pictures 166 top, 268.

Photolibrary/The British Library 102; /Nicolas Thibaut 146 centre.

Scala 251; /White Images 150.

Science & Society Picture Library/Science Museum 49, 55 bottom right, 85 bottom right.

Science Photo Library/13, 19 top, 29, 40, 47, 54 top centre, 68, 70, 75, 81, 84, 101 top right, 101 top left, 104 top, 117, 122, 126, 130, 142 top centre, 143 top left, 143 top centre right, 143 top right, 147, 156 right, 161, 185 top left, 192, 193 top, 195 bottom, 196, 201, 210 right, 212 top, 216, 218, 221, 228 top right, 229 bottom right, 244 bottom, 246, 247, 256 top, 256 bottom, 261 top, 265 left, 265 right, 274 top right, 275 top left, 275 top right, 60 top, 78 top, 181 top, 191 right; /Steve Allen 54 background picture 5, 100 background picture 5, 142 background picture 5, 184 background picture 5, 228 background picture 5, 274 background picture 5; /American Institute of Physics 46, 91 top; /Anatomical

Travelogue 205 left, 205 right; /A. Barrington Brown 222; /David Becker 264; /Juergen Berger 209; /George Bernard 11, 111, 114 right, 143 top centre, 144; /Paul Biddle & Tim Malyon 189 bottom; /Maximilien Brice, CERN 56; /Dr Jeremy Burgess 191 left, 242; /Caltech Archives 34; /CCI Archives 190 right, 228 top centre right; /J-L Charmet 168 top; /Russell Croman 152; /Custom Medical Stock Photo 213 top; /Dopamine 206; /John Durham 210; /Bernhard Edmaier 123; /Emilio Segre Visual Archives/ American Institute of Physics 55 top centre right, 96, 101 top centre left; /Equinox Graphics 134; /Prof. Peter Fowler 92, 101 top centre right; /Simon Fraser 200; /General Research Division/New York Public Library 118; /Steve Gschmeissner 226; /Tony Hallas 52; /Roger Harris 240; /Adam Hart-Davis 208 top right, 208 left, 229 top left; /Gary Hincks 136; /Keith Kent 155; /Gavin Kingcome 110 top; /James King-Holmes 2; /Mehau Kulyk 188, 190 left, 249; /Andrew Lambert Photography 80; /Patrick Landmann 135; /Library of Congress 146 top, 167, 172 top, 178; /Library of Congress/New York Public Library 166 bottom; /Living Art Enterprises, LLC 91; /Medical RF.com 270; /Astrid & Hanns-Frieder Michler 91 bottom; /Mid-Manhattan Picture Collection/Glass/New York Public Library 169; /Cordelia Molloy 59 bottom, 173, 179 bottom, 185 bottom centre left, 185 bottom centre right; /NASA 6, 8 top, 39, 48 centre; /NASA/ESA/STSCI/L. Sromovsky, UW-Madison 48 left; /National Library of Medicine 258; /NOAA 33; /Omikron 277; /David Parker 99, 276–277; /Pekka Parviainen 28; /Pasieka 72, 216, 229 centre, 244 top; /Photo Researchers 179 top, 185 top right, 208 bottom right, 228 bottom right, 229 top right, 274 top centre left; /Physics Today Collection/American Institute of Physics 90; /Philippe Plailly/Eurelios 140; /Maria Platt-Evans 108 top, 142 top right, 158 top, 185 top centre left; /Radiation Protection Division/Health Protection Agency 180; /Detlev Van Ravenswaay 19 bottom, 26 bottom left, 54 top left, 55 bottom left; /Royal Astronomical Society 21, 26 centre, 45, 55 background picture 2, 100 background picture 8, 142 background centre, 184 background centre , 229 background picture 1, 274 background centre; /Friedrich Saurer 32; /Jon Stokes 260 left, 275 bottom left; /Science, Industry & Business Library/New York Public Library 168 bottom; /Science Source 59 top, 64 top, 100 top left, 204, 220,

229 top centre right, 260 right; /Sovereign, ISM 230; /Sinclair Stammers 110 bottom; /St. Mary's Hospital Medical School 14; /Mark Sykes 64 bottom; /Sheila Terry 25, 36, 44, 54 top right, 55 top left, 55 top centre left, 78 bottom, 79, 79 bottom, 100 top centre, 100 top right, 116, 119, 129, 133, 143 top centre left, 171, 174 top, 182, 184 left, 185 top centre, 185 top centre right, 188 bottom, 188 top, 189 top, 194, 195 top, 197, 228 top left, 228 top centre left, 229 top centre left, 235, 237 bottom, 242 right, 250 right, 274 top centre, 274 top centre right, 274 bottom left; /G. Tomsich 193 bottom; /Michael W. Tweedie 227; /US Department of Energy 101 bottom right; /US Library of Congress 51, 55 top right; /US National Library of Medicine 101 background picture 1, 143 background picture 1, 185 background picture 1, 229 background picture 2, 261 Bottom, 275 background picture 1; /Charles D. Winters 76; /Ed Young 98.

The Natural History Museum, London 104 bottom, 105, 114 left, 142 top left, 142 bottom left.

TopFoto 169 left; /Fortean 24; /The Granger Collection 38, 93.

Wellcome Library, London 237 top, 248, 253, 252, 254.

De Beer Collection, Special Collections, University of Otago, Dunedin, New Zealand 126, 143 bottom left.

Richard Wheeler, Sir William Dunn School of Pathology, University of Oxford 224–225.

ACKNOWLEDGEMENTS

The television series on which this book is based is presented by Michael and executive produced by John, for whom a television history of science has long been a goal. The concept of the series has its origins in the development team of BBC Science, particularly Ros Homan who set out the framing concept of the six great questions. The programmes themselves were shaped and crafted under the inspiring editorial leadership of series producer Aidan Laverty. The research team of Alice Jones, Naomi Law, Liz Vancura and Will Ellerby took on what is effectively a history of everything to search out compelling stories, which were brought to life by directors Jeremy Turner, Nat Sharman, Peter Oxley, Nicola Cook, Giles Harrison and Nigel Walk, while the sprawling production process was managed with true dedication by Maria Caramelo and Sarah Forster, under the ever-wise guidance of production manager Giselle Corbett.

Countless experts and contributors gave time and effort to the preparation of the television series, but none more so than its three highly-committed academic advisers, Pietro Corsi, Jim Endersby and Patricia Fara, whose expertise, wisdom and insight guided the production team's path through the programme scripts.

For the book, we are indebted to the commissioning efforts of Peter Taylor, the editorial guidance of Georgina Atsiaris and the design talents of Pene Parker, Yasia Williams-Leedham and Mark Kan. Special credit goes to Hayley Birch for her skills in smoothing the rough edges of our draft chapters, for integrating our often differing literary styles and for keeping a weather eye open for embarrassing mistakes - although we retain responsibility for any errors that remain. A special thanks also from Michael to Ewan Fletcher for his help – and from John to DM Lawrence for her unflagging assistance in managing the day.

Finally, it is simply not possible to undertake a challenging project like this without the continued support of friends and family. Michael could not do what he does without Clare; and John has been humoured by Ewa, Christopher and Toby over the course of a difficult summer and autumn. We hope they all feel it was worth it.